UNITED NATIONS RESEARCH INSTITUTE
FOR SOCIAL DEVELOPMENT

SEEDS OF PLENTY, SEEDS OF WANT

Social and Economic Implications of the Green Revolution

by
ANDREW PEARSE

CLARENDON PRESS · OXFORD
1980

Oxford University Press, Walton Street, Oxford OX2 6DP

London Glasgow New York Toronto
Delhi Bombay Calcutta Madras Karachi
Kuala Lumpur Singapore Hong Kong Tokyo
Nairobi Dar Es Salaam Cape Town
Melbourne Wellington

and associate companies in
Beirut Berlin Ibadan Mexico City

Published in the United States
by Oxford University Press, New York

British Library Cataloguing in Publication Data

Pearse, Andrew
    Seeds of plenty, seeds of want.
    1. Green Revolution
    I. Title   II. United Nations. *Research Institute
    for Social Development*
    338.1′6   HD1417   80-40469
    ISBN 0-19-877150-9

Typesetting by Hope Services, Abingdon, Oxon
Printed in Great Britain by
Billing & Sons Ltd., Guildford, London and Worcester

AUTHOR'S NOTE

The argument of this book is divided into three parts. Part I (Chapters I and II) deals with some essentials of the conceptual framework, and the characteristics of the particular technology and its propagation. Part II (Chapters III to VIII) is based on an assumption that variations in agrarian structure directly affect the mode of technology-induced economic and social transformation. Three types of agrarian structure are introduced: (i) structures found mainly in Africa south of the Sahara and in some less populated areas in Asia in which traditional forms of communal tenure are still widespread; (ii) 'bi-modal' agrarian structures in which the larger property directly farmed by subjected labour or wage labour is set amongst settlements of peasant agriculture and in which an unequal but symbiotic relation exists between them. This form has predominated in Latin America and North Africa; (iii) Asian peasant structures which (with important exceptions) are now deeply penetrated by extortionate tenancies. Three other related questions take a chapter each, namely: State-backed programmes of technological change, the size of holdings, and the effects of mechanization, especially on employment. Part III (Chapters IX–XIII) begins with a summing-up of the 'critical issues' that have arisen with the large-scale introduction of the genetic-chemical technology and the accompanying mechanization and capitalization of agriculture. Three further chapters take up questions about economic and technological policies that avoid the generation of poverty. The last chapter is about 'peasant-based' strategies in differing social systems by which productivities are increased without gross polarization of wealth, and the examples of Japan, China, and Taiwan are cited.

October 1979
Queen Elizabeth House
Oxford   ANDREW PEARSE

ACKNOWLEDGEMENTS

In addition to the names of all those whose contributions are referred to as authors of reports, articles, and books, I wish to acknowledge those who helped with the final version through editorial work and comments including Solon Barraclough, Antonio Barreto, Bridget Dommen, the late Erich Jacoby, Joyce Maxey, Marc Nerfin, Wendy Salvo (née Beale), Pierre Spitz and W. F. Wertheim.

I also wish to mention the efforts of Stig Andersen, William Mashler, and Donald McGranahan, who steered the courses respectively of the three United Nations agencies most directly involved, namely the Office of Technical Co-operation and the Development Programme (UNDP) both of New York, and the Research Institute for Social Development in Geneva.

Finally, David Millwood deserves abundant thanks for the way in which he cleaned up the manuscript and saw it through to the Press.

PREFACE

UNRISD's Green Revolution studies have shown that two of the leading features in the crises of livelihood in most of the developing world are (i) the emergence of more capital-intensive, higher technology farming and (ii) the accelerating dissolution of self-provisioning agriculture, both as a major element in peasant farming and as a subsistence base for the poorer rural strata.

The commercialization of production and exchange relations, the growing competition for good quality lands by entrepreneurial farms, and the increasing numbers of landless labourers and of families trying to extract a living from diminishing areas of poor quality lands all contribute to this process of decay. The food systems that have maintained humankind throughout most of its history are disintegrating before other forms of economic activity are able to offer alternative means of livelihood to the displaced peasantry.

The full significance of this transformation is not entirely comprehended, but it seems to imply deterioration in the nourishment of the already poor obliged to purchase food in unfavourable conditions from the market, massive migration to urban centres, growing unemployment and underemployment, and a much higher level of conflict, disorder, and repression.

These contradictory trends affecting negatively many social groups can in no way be considered to have been 'caused' by the new high-yielding varieties of food grains. Rather they are a product of complex interrelationships among and within social systems that happen to have received in some instances additional perturbation from the introduction of the new varieties. This book makes clearer the nature and complexities of these social processes.

The origins of this book go back to 1970, to the decision to undertake a wide examination of the socio-economic effects of the Green Revolution. The series of studies known as Global Two was a joint undertaking by the United Nations Research Institute for Social Development (UNRISD), which commissioned it, and the United Nations Development Programme (UNDP), which contributed to financing it. The multi-disciplinary team that conducted the studies was led by Andrew Pearse, the author of this book.

Much of the data on which *Seeds of Plenty, Seeds of Want* is based has already been published and has had a major influence. (Details will be found in the book list.) A Summary and Conclusions was prepared (in 1974) and a draft Overview Report (1976) was circulated for comments but was not published.

What seemed to be needed in order that the important results obtained from this enterprise reach as wide an audience as possible was a more analytical and interpretative summary of the conclusions, one that presented clearly the main policy issues with their political implications.

Although this meant a further delay in publishing the conclusions of the Green Revolution studies, I believe that *Seeds of Plenty, Seeds of Want* fully justifies that decision.

A companion volume, consisting of village level case studies from the Overview Report, is in preparation and will be published shortly.

SOLON L. BARRACLOUGH
Director, UNRISD

CONTENTS

INTRODUCTION

This book is about the social and economic implications of the so-called Green Revolution.[1] Its central task is to bring into focus the relation between modern capability in agricultural technology and adequate feeding for the world's peoples. This is approached by showing what happens when the new technology is introduced into different societies, so as to help clarify the issues and problems, especially for those who are involved in making and carrying out policies to assist rural people, dependent on agriculture, to achieve adequate diets and assured livelihood.

We shall, therefore, be discussing a number of questions that lie at the very heart of the problem of food and also some of the ways that the problem itself is perceived, for example by governments, by technicians and researchers, and by those people on whom the success of any policy or programme ultimately depends: the cultivators themselves.

The Green Revolution was an international campaign aimed at increasing the productivity of land by means of the introduction of a science-based technology (referred to throughout for convenience rather than for accuracy as 'the new technology') in the production of foodgrains. The campaign succeeded in bringing about the introduction of the technology on a sufficient scale to give it great socio-economic significance and to merit the closest inspection as to its implication: hence the study known as Global Two sponsored by the United Nations Development Programme and carried out between 1970 and 1974 by the UN Research Institute for Social Development, on which this book is based.

Essentially, three related aspects of the process of technological change in the agricultural/rural sector were studied. These were:

(i) identification of the factors facilitating or obstructing

[1] As will be seen, the main sponsors of the campaign to implement the technology shared a common doctrine of development strategy, so that the catch-phrase 'Green Revolution' came to mean the technology plus the strategy, expressed in country programmes and international research policy.

the acquisition and use of a 'genetic-chemical technology';
(ii) identification of the economic and social changes that
follow the large-scale introduction of the technology to
be observed in the agrarian structure, in the level and
quality of livelihood of the participants, and in the social
structure of the rural society;
(iii) assessment of measures and programmes proposed and
carried out by governments in order to manage or modify
the processes set off by technological change.

Research Guidelines

Studies were contracted by agronomists, sociologists, agricultural economists, anthropologists, political scientists, and
historians. They sought to illuminate and explain the emergent situations at different levels and in different ways. An
inventory of what seemed to be significant issues was provided
but the researchers nevertheless gave the characteristic angle
to their individual studies according to their sensitivity to
local issues, their own situation in the society, and their
respective disciplines.

Each used his or her own methods of research. No standardized instruments were insisted upon but it was made clear
that field studies were required that would approach the
livelihood and social relations of the different groups and
classes of persons involved in agricultural production, the
functioning of agrarian institutions, the operation of units of
production, and forms of marketing and exchange at the
local level as well as the real performance of governments,
private sector agencies, and purveyors of the elements of the
technology in question.

Most of the studies that resulted attempted to look at the
introduction of the new technology in the context of one or
several primary rural settlements or, better still, in the context
of such settlements and their relations with the urban market
centres on which they were partially dependent.

A number of studies, however, used data gathered by
sample surveys of productive units within a more extensive
area. While these studies provided aggregate data comparable
with regional and nationwide statistics, they were seldom

able to provide explanations of the persistence or alteration of the form and content of social and economic behaviour, which, however, becomes evident in the framework of community relations and in the specificity of the locality at a particular point in its history.

Objectives and Assumptions

The evaluation of policies and programmes of course required clearly established and agreed notions about what the objectives of the introduction of the new technology were. This key problem was finally solved by taking the position that governments, scientific institutions, and technical assistance or development agencies were able to give their assent to the proposition that the two main objectives of the technological changes contemplated were:

(i) Freedom of nations from food dependence through accelerated increases in food production leading toward food self-sufficiency; and

(ii) Freedom from hunger for their populations.

While the prime importance of these two objectives is generally agreed, the relation they have to one another, and the order of priorities and means of realization, are the subject of disagreement and conflict.

In United Nations circles there was ambiguity over the assumptions underlying the study, especially in regard to its evaluative character and the nature of policy recommendations that should arise out of it. Certain voices could be heard urging that the general policy of the Green Revolution should not be questioned but that research should identify any undesirable social consequences and, it was hoped, put forward some practical measures for countering these. Other voices favoured a less restricted approach.

The propriety of making policy recommendations to the governments of member states was also held in question, and there was an understandable sentiment among the social scientists who were responsible for the research that a statement of recommendations would be out of place.

The advisory committee for the project agreed that the research should lead to a report to be presented to governments, accompanied by a text that might be paraphrased as follows:

Sirs, if you are about to embark on discussions about your agricultural development policy, and especially if you are interested in finding a land-saving cereal crop technology, we pray you to scan this report first since it gives an account of what took place in a number of countries, some of which bear a resemblance to your own. It may point to some of the issues that require your special attention.

It was in this spirit that the study was carried out but, nevertheless, by the time the field work was completed two years later, it was assumed by the sponsors that policy recommendations, even if they were to be presented under the title 'Policy Alternatives', were a required product of the research. And indeed by this time those who had set out on a task involving analysis and explanation had reached firmer convictions that they felt had acquired some authority.

A 50-page document[2] finally emerged presenting a series of conclusions arising from the field research and the analysis of other studies and public statistics. It was published in Geneva in 1974, and at the same time it was circulated to member states by the UN Secretariat, with an introduction by the Administrator of the United Nations Development Programme. This document consisted of three parts—the first dealt with conditions and constraints affecting the introduction of the new technology, the second with changes at the local level associated with its adoption, and the third with policy implications. Its contents and language go rather beyond a sort of Highest Common Factor of the various interests involved, in the direction of an area of policy consensus of the researchers. Although it was circulated to all member states, probably to their Planning Offices and Ministries of Agriculture, the author is not aware of any feedback at the national level, and can make no comment on its impact.

Sixteen reports were published and are listed by author in the booklist. The present volume is an overview of the whole

    [2] *The Social and Economic Implications of Large-Scale Introduction of New Varieties of Foodgrain: Summary of Conclusions of a Global Research Project*, UNRISD Report No. 74.1, Geneva 1974.

research and will be followed by an additional volume containing a selection of grass-roots 'vignettes', studies at the level of the rural locality.

The whole problem about a discussion of 'policy implications' is of course that the range of feasible alternatives open to governments is limited by the interests of the sectors supporting them, and what may be feasible in one national situation can only be described as utopian in another. Nevertheless, utopian or not, the United Nations is identified with a commitment to freedom from hunger, and this fact establishes an obligation to propose a strategy that would ensure not only adequate production for over-all needs, but a pattern of distribution that would ensure to all families a minimum of food necessary for health.

In fact, the achievement of basic food security by the poor in a market-centred post-colonial economy would require a most difficult kind of transformation. Not surprisingly, our studies reveal the prevalence of tendencies leading in the opposite direction as new technology and facilities are injected into agricultural societies already dominated by excessive inequalities and debt. We have named these tendencies the *talents-effect* after the well known Biblical parable in which it is recounted that one servant receives money to the value of ten talents from his master and is able to invest and prosper, while the very insecurity of his humbler fellow restrains him from utilising the single talent entrusted to him, which is wrathfully reappropriated by the master, and given to the successful investor.

> 'For unto everyone that hath shall be given
> and he shall have abundance: but from him
> that hath not shall be taken away even
> that which he hath'
> (Matthew, Ch. 25, v.29)

The issues involved in this contradiction are central to the discussion of policies and strategies in the latter chapters.

PART I

I THE NEW TECHNOLOGY AND THE PEASANTS

*The theme of the book is outlined and the problems of
food production looked at from the viewpoint of both
government and the cultivator. The motivation of the
working cultivator and his dilemma in the face of rapid
technological change and market incorporation are
introduced.*

Technology and Society

In contrast to the belief that technology can be neutral, our
suppositions for the Global Two studies were based on the
belief that technologies and social relations are intimately
linked and that a change in technology was virtually always
associated with a major change in social relations.

Many cases of this association are provided by the history
of agriculture. The clearance of the Scottish Highlands and
the large-scale migration of Highlanders across the Atlantic
was the immediate result of the discovery that by introducing
the Cheviot sheep in place of deer, cattle, and valley cultivation, extensive agriculture could become a highly profitable
investment:

So began the invasion of the Cheviot or True Mountain breed. They
came up the old cattle roads into Argyll, Inverness and Ross. They
climbed where the deer died, they throve where the black cattle starved.
Land which had produced two pence an acre under cattle now yielded
two shillings under sheep. Four shepherds, their dogs and three thousand
sheep now occupied land that had once supported five townships. Small
gentry, lawyers, merchants, half-pay officers with a little prize-money
took up leases on land they rarely saw and became Highland gentlemen
with imaginary pipers and gillies at their tail. (*Prebble, 1963*, p. 28.)

These changes made it possible to establish capitalistic farming enterprises paying heavy rents to landlords who had formerly been clan chiefs.

Similar structural changes, involving the destruction of small-scale agriculture producing food crops for the local market, resulted from the improvements in sugar technology, both at the level of processing in mills and factories and at the level of productivity in the field. This has been reported on from North-East Brazil and Cuba by Correia de Andrade and Du Moulin (*Pearse, 1975*) and from Colombia (*Taussig, 1971*). The harassment and effective liquidation of small-scale agriculture in the fertile Valle del Cauca, Colombia, as a direct result of the great advances in production technology not only for sugar but for other commercial crops, forms the theme of one of the Global Two studies (*Rivera, 1972*).

The widespread introduction of a genetic-chemical technology is inseparably linked to major structural changes in the society and it is therefore unwise to talk about 'technological neutrality' in this connection. In another sense, however, a technology *is* neutral in that the significance of its use depends on the purposes and situation of the user. In theory at least, governments are able to make the new technology serve their chosen development policy. In fact, the social forces upon which governments rest set parameters for technological policy.

There is a certain amount of ambivalence in the phrase 'Green Revolution' since it is used both for a technology and for a certain kind of strategy used to implant it. The present volume is about the Green Revolution but the phrase is avoided in any precise sense, although the idea of a Green Revolution strategy is used. In referring to the technology apart from the strategy, the phrase 'new technology' is used, with due apology for its inaccuracy.

What Is The New Technology?

Attempts to expand the production of food crops in human history have usually followed rising population trends or urbanization, and the most common method is to extend the area of land used. However, where land area is already limited, production may be intensified, and yields raised by modifications or transformations of productive technology have been made by countries with limited and finite supplies of agricultural land, in which industrial development has been drawing an increasing proportion of the rural population into manufacturing and tertiary activities in cities, and at the same time has provided a basis for research advances applicable to agriculture.

The highest average yields for foodgrain in 1970 were to be found in the Netherlands, United Kingdom, and Belgium, at 4,460, 4,130, and 3,890 kg. of wheat per hectare respectively, while Japan and Taiwan led the field in those countries where rice was the main staple diet, with 5,640, and 3,990 kg. respectively.

The improvement of foodgrain yields by means of seed selection and cross fertilization has a long history, but many of the elements of the 'genetic-chemical' technology with which we are concerned are the result of research and development efforts of the last few decades. A number of these basic scientific developments were listed by FAO and are given in Table 1.

TABLE 1

*Some Significant Innovations in Crop Production
and Dates of Widespread Use*

Hybrid maize	1933
Chlorinated hydrocarbons for insect control	1945
Minimum tillage	1945
Foliar feeding	1945
Direct applications of anhydrous ammonia	1947
Chemical weed control	1951
Systematic biocides	1953
Hybrid sorghum	1957
Dwarf wheat	1961
Dwarf rice	1965
Opaque 2 maize (high lysine)	1965
Hybrid barley	1969
Hybrid cotton	1970

The éclat and applause that accompanied the Green Revolution was not due primarily to the originality of its discoveries nor to the high level of yields achieved, since these in most cases fell short of yields in industrial countries, especially those with limited land area. Its successes were rather in the elaboration and adaptation of existing technology to the needs of typical environments. The release of these new strains was the occasion for a simultaneous rush by a number of land-poor Third World countries to close as rapidly as possible the technological gap that had hitherto separated the tropical colonies from the metropolitan powers.

With one or two exceptions, such as Taiwan (then Formosa) in relation to Japan, the colonial territories had been prized by their metropolitan masters for commercial crops to which yield-improving research had been devoted. 'Native agriculture' had been allowed to make its own way so long as labour was available for the production of tropical commodities for the international market.

The new technology is built around the use of man-made varieties of wheat, maize, and other foodgrains in man-controlled environments. The qualification 'man-made' refers to the fact that the new varieties are the result of a sophisticated scientific process of selective breeding designed to result in stable new varieties of plant having certain desired qualities, most of which are favourable to higher yields and higher production. The basic biological characteristics of High-Yielding Varieties (HYVs) has been their potential responsiveness to plant nutrients from both soil and the action of sunlight through photosynthesis.

But increased nutrient-uptake required greater structural strength. Dwarf and semi-dwarf plants meant that more of the nutrients went into the production of the grain itself, whilst the stockier, more robust build of the shorter plant enabled it to support heavier grain clusters (panicles) without 'lodging' (collapsing) and to withstand high winds or rough treatment at the hands of reapers, whether human or mechanical. Thus dwarf and semi-dwarf wheat and rice can be given much larger doses of artificial fertilizer without running the risk of loss as the slender local varieties often do.

A high tillering rate is obviously necessary since the individual cereal plant enlarges its fruitfulness by sending up from the roots shoots or sub-stems known as 'tillers', with a capacity to form grain-bearing panicles or ears, and also, of course, to produce leaves that contribute to growth by photosynthesis, provided they are not overcrowded. Photosynthesis —the combining of carbon dioxide in the air with water by the action of sunlight to form carbohydrates—is a normal and essential part of the process of growth. But it has been shown by experiment that the uptake and utilization of very large dosages of nitrates requires increased solar radiation, so that the maximization of leaf surface exposed to sunlight, and the setting of the leaves in relation to the sun's trajectory, become a necessary complement of high-level fertilizer application. So a high tillerating rate and a tendency for the leaves to have an erect position are both important characteristics of HYVs.

The photosensitivity of the plant is also important in connection with the timing of the vegetative cycle and its synchronization with seasonal weather conditions. Non-photosensitivity has been bred into plants so that they will mature within a given period, unaffected by annual variation in the climate or by local variation in day-length, especially to ensure a rhythm of work making multi-cropping possible. However, in other conditions it may be important to follow the irregularities of nature; for instance, in some areas, it is difficult to harvest rice before the soil is dry, so photosensitivity is a desirable feature since in case of a late or especially heavy monsoon, the maturity of the plant and the necessity of harvesting it are delayed.

Still other qualities the scientists attempt to control have to do with the intrinsic value of the plant to consumers, such as its nutritive and culinary qualities, its taste, and its colour. Improved maize and sorghum varieties differ from the other main foodgrains in their method of reproduction. This may be open-pollinated, which is less productive but with permanent genetic features allowing it to be reproduced indefinitely from the previous generation, or hybrid, which is exceptionally productive for the first planting only, making yearly renewal of seed necessary. With maize, a third group of 'synthetic' varieties has been obtained by combining selected lines of plant and reproducing and multiplying them for four to five generations by uncontrolled pollination. Yields are usually between those of the other two types, but the seeds need to be replaced only every three or four years.

The man-controlled environments in which these more delicate custom-built plants can best thrive are achieved by means of the use of manufactured chemicals, carefully timed and rationed applications of moisture and appropriate cultivation recipes. The function of the chemicals is to increase available plant nutrient to the optimum level of capacity for these more voracious varieties; to control the onslaughts of pests and disease, intensified in response to the increase in host material, and to destroy weeds that also luxuriate in the surfeit of fertilizer.

Cultivation practices themselves are of course a way of controlling the environment, but the high potential yield of which the new grains are capable can be achieved only if such practices as weeding, watering, fertilizing, transplanting, and plant-spacing are all carried out in a specially stipulated manner, which is more demanding of accuracy and labour than customary husbandry.

Given the number of variables within the seeds themselves and in the inputs and environment on which they depend, as well as the varying requirements of the growers, it cannot be surprising that the success of HYV techniques has been extremely uneven. Even when the so-called high-yielding seeds are accompanied by improved practices and inputs, there have been occasions when their performance has not been superior, and sometimes has been inferior, to locally improved or traditional varieties (though this has been occurring less frequently as HYVs have been crossed more and more with traditional varieties). There have been cases, too, where yields have been raised by using more fertilizer without changing to the new seeds or significantly increasing other inputs. Indian trials on rice plants between 1958 and 1961— before HYVs were available—showed that in most parts of India a fertilizer increase of up to 22 kg./ha. would have produced a profitable increase in the yield of traditional seeds.

Perhaps the crux of the technical problem lies in obtaining a balance between all the elements in the production process that is an optimum match for local conditions. It is on account of the complexity of this problem that sponsors of the technology developed the idea of a 'technological package'.

The 'Packaged' Technology

Let us now consider some of the problems connected with the implantation of a new exotic agricultural technology in a milieu where hitherto cultivators have used locally evolved and adapted practices and plant varieties. The idea of a package is made central in this discussion and it is used to mean a set of recommended tools, material inputs, and practices—somewhat like a cook's recipe. Although the concept *package* is used widely by scientists and programme officers, it rather misleadingly suggests that all the necessary elements for cultivation can be made available together to the cultivator as a single 'package deal', but reality is far different from this.

Programmes vary in the manner and extent to which they set up facilities and arrange delivery systems for inputs, but in most of the cases that came within range of our study, only in numbered instances could the cultivator have the impression of a balanced and easily accessible package. In most cases, he had to make arrangements as best he could with several different public and private sources—in other words, the package existed as a norm or model for the programme officials, *but its realization depended on the cultivator* and the kind of service relation he could establish with these sources.

The essential requirements of the package as a normative recipe or pattern fall mainly under two headings: that of fitting the package to the existing physical conditions (viz. agronomic problems) and that of ensuring that the product can be marketed on terms more advantageous to the cultivator than those on which the current product is marketed. This second prerequisite, an economic one, is more variable and elusive.

As regards the first, agronomic suitability, it is very important to realize the difficulty of choosing between various options in laying down a plant-breeding programme, the results of which may only become operative after a few years.

The demand for the new product, which in turn affects the price, is influenced by such features as colour, consistency, taste, and smell and other culinary characteristics, and also by the degree of sensitivity of the plant to varying environmental conditions, which affect the risk involved in planting it. Of course, other very important factors, beyond the bounds of simple prediction, affect the situation. When the present study was initiated (1971), fertilizer prices showed a tendency to decline. At the time of writing, the prices of chemical products both for fertilization and for plant protection have risen so high that they threaten the continued use of the high-yielding varieties for those small cultivators for whom they furnish no more than marginal advantage over local varieties. Thus the planning of future packages requires a prolonged study of trends likely to mould the future situation. Quite apart from special contingencies, the fitting of a cultivation- and production-recipe to specific natural conditions can be a complex affair.

At best, in physically homogeneous conditions, the sponsors of the package may have to decide between the technological optimum, i.e. the amount of fertilizer giving the highest levels of production, and the economic optimum for the cultivator, i.e. the amount rendering the highest net gain. But in areas of great physical variation the choices before the sponsors are infinitely more complicated. Agronomists working in the Eastern Cundinamarca Project, Colombia (*Zandstra, 1973*) found that within a single restricted zone (Caqueza and Chipaque) of that mountainous area, there were wide climatic differences and also great differences in phosphorus found in the soil, which influenced the amount of the chemical that should be included in the package.[1]

---

[1] At one extreme, it was possible to make a single recommendation for the whole region based on an average drawn from tests made at different points and experiments in yield-response to applications of graded amounts of fertilizer. This method promised administrative simplicity and low cost. Yet cultivators whose fields contained a high existing level of phosphorus and those that contained none at all would both suffer losses, the one through expenditure on unnecessary phosphorus and the other through low yields resulting from inadequate applications of phosphorus. And extended patches of unprofitable results from the package might create significant localized scepticism. At the other extreme, every cultivator might be called upon to have his soils submitted to laboratory analysis. This would make possible made-to-measure packages for every cultivator, a highly satisfactory situation but an expensive and not very feasible operation in a smallholding area.

Between these two extremes lie a number of intermediate options. For instance,

The complications of the operation of choosing between these various approaches to the determination of the package is to be seen in Table 2 with an attempt at costing the different options.

One of the lessons learnt from the problems revealed in the search for the 'right package' in the Colombian Andes has to do with the relationship between technical staff and cultivators. While the former are able to interpret the potentialities of the new seeds, chemicals, and techniques, the latter are likely to know pragmatically, but often subtly, how their land and the local climate can be expected to behave. With the increasing importance of chemical fertilizers, the cultivator requires to know what phosphorus does to his land and its plant life, and which of his observations in his own fields can be translated into reliable indications of the presence or absence of the desired elements. The perspective for the cultivator taking up a science-based technology must be his ever-increasing knowledge of scientific principles and their application to his productive activities.

In certain respects, the case discussed above is exceptional on account of the characteristic soil/climate varieties found in mountainous territory. But it brings into focus a problem of approach that recurs throughout the present work. Zandstra and his colleagues, who investigated the problems of arriving at a recommendable package, concluded that the only socially feasible approach was one that set out from existing peasant practice and concentrated on solving problems of production as perceived by the peasant cultivator. Their approach contrasts sharply with the more usual practice of

the cultivator may be asked to bring a soil sample with him when he applies for credit and comes to buy his fertilizer. This sample could be submitted on the spot to a test much simpler than a laboratory analysis, in order to classify it as 'high' or 'low' in phosphorous and thus qualify him for 'Package A' or 'Package B'. This procedure also has its costs, but they are lower than those of individual soil analysis.

Another possibility consists of variations according to altitude. In the case reported in Colombia, however, one of the *raisons d'être* for the high phosphorus patches was earlier use of the land for potatoes or for garden crops and the consequent existence of important residual levels of phosphorous applied to these crops. A further approach, subsequently implemented, was to classify the land on the basis of the history of its use as reported by the farmer, and to prescribe accordingly.

TABLE 2

*A Comparison Between Different Approaches to Recommending Quantities of Phosphorus (P)
for Maize in the Municipalities of Caqueza and Chipaque in Cundinamarca, Colombia*

Approach to assessment of fertilizer needs	Addition to net profit per ha.	Cost of the approach, i.e. test, analysis etc.	kg. of P/ha.	Net benefit anticipated	Anticipated benefit to region	
					Caqueza and Chipaque	Eastern Cundina-marca
	Pesos				Pesos '000	Pesos '000
A single prescription for all maize fields	189	0	34.0	189	404	1,108
Potato localities 0 kg./ha. Maize localities of 53 kg./ha.	189	0	—	189	404	1,108
Two prescriptions, fitting 'high p' and 'low p' classification by test	344	50	28.3	296	633	1,735
Three prescriptions instead of two as above	430	50	—	380	813	2,228
Prescriptions for each farm on the basis of individual soil analysis	488	110	10.1	338	723	1,982

Source: Zandstra, 1973, p. 33A.

rural development programmes, which is dominated by the belief in the capacity of scientific and technological research to find the right way and to teach it to the peasantry. Indeed, it is possible to go further and to show the discrepancy between the 'problem' to which official programmes involving governments, planners, scientists, and extension officers address themselves, and the 'problem' to the solution of which the common cultivator bends his labours and upon which he concentrates his ingenuity and received wisdom.

The 'problem' of the official programmes is usually one of national production figures, and the bearing of these on the price of common food for more organized and vocal sections of the population (viz. the urban population). National food self-sufficiency is also sought to bring relief from undue pressure on the balance of payments and to escape from excessive political dependence on food exporting and 'aid-giving' countries.

The ideology that accompanies this view of the 'problem' encourages an image of the agricultural sector in which the rational and scientific solutions of the problems of production are constantly being generated by the scientists, but are not 'adopted' by a backward peasantry. It is assumed that the reasons for the adoption or non-adoption of the practices are to be found in the ignorance and conservatism of the peasants. In some popular lines of research, cultivators are classified as 'progressive', 'early adopters', even 'back-sliders', as if elements of technology were articles of religious faith and their behaviour attributed to personality traits common to peasantries, such as fatalism or cosmopolitanism.

This view of the matter is liable to support discriminatory action by the extension agency, which will seek production successes by pushing ahead with a number of the larger farmers whose circumstances make the innovations a rational step forward (and giving them a claim to be classified 'progressive') while turning away from the majority or minority whose circumstances forbid or inhibit the step.

This attitude is rightly criticized by Havens and Flinn (1975, p. 18) who consider that rather than concentrate on 'the problem of the backward peasant', who cannot keep up with advance in technology, who is resistant to change, and so on, research should be a critique of the respective societies and their institutions, and their failure to provide circumstances and facilities enabling the small cultivator to embody the fruits of scientific research in his system of production.[2]

The Pursuit of Livelihood

While rejecting the notion that slowness in applying technological innovations is to be explained mainly by peasant traditionalism, backwardness, and other negative attributes, we do not imply that policies can be based on the expectation that all kinds of cultivators will be able to respond in a 'business-like' manner to price signals. Indeed, the most realistic assumption on which to base an explanation of the motives and decisions of cultivators is that they all pursue livelihood and that their decisions about technological practices must submit to the exigencies of the tactics of this pursuit.

The concept of 'livelihood' is therefore at the centre of the motivational scheme. The cultivator's actions may be thought of as addressed to the achievement of a certain mode or norm of livelihood.

The suitability of the word 'livelihood' for conceptualizing 'means of living' in pre-capitalist conditions is underlined by the very loss of meaning, the vagueness of the word in contemporary usage. In our industrial societies, money is the universal means to the increasingly standardized and commercialized goods and services used. There is justification, therefore, in restoring to 'livelihood' its earlier concreteness for use in discussions about less commercialized societies, and using it to denote *that quantum of goods, services, and facilities needed to maintain the life of an individual or family*.

Reference to the livelihood of a family is not simply a way of talking about their standard or level of living, nor do we intend to use it as a vague reference to welfare. A real livelihood may fall far short of the image or model of livelihood that

[2] These authors, in their work for Global Two, mention as constraints the structure of land tenure, lack of political participation, economic segregation, and the inequitable distribution of wealth, of services, and of legal privileges and rights. They further criticize the failure of researchers to take into account the system of advantage and handicap inherent in class hierarchies, and the facile unsound assumption implicit in certain adoption models that all cultivators enjoy equal access to public institutions dispensing services and information.

the family and its members aspire to and pursue as a goal. Both as model and as quantum consumed and used, livelihood varies with the variations in status within a particular community, society, and culture.

This approach does not provide immediate explanations. These only emerge from a much fuller understanding of the way in which the struggle for livelihood is conducted. But it puts the enquiry on the right track. At one level, the entrepreneurial farmer is found to pursue financial gain by means of which livelihood can be bought with cash or on credit. He is likely to be a 'profit maximizer' and his productive activity is also likely to be conducted by means of the payment of money wages to labour. He purchases inputs and sells his product for cash and credit, backed by the social, economic, and political power likely to accompany his entrepreneurial activities in a developing capitalist society.

At another level is the cultivator who regularly produces a commercial crop, relying upon income from its sale for the purchase of necessities and other items of consumption. He is accustomed to rely upon the market for a varying but substantial part of his livelihood, with support from his own domestic production.

A third typical cultivator puts greater reliance on his ability to produce his own essential needs, with the sale of a certain portion of the product surpluses providing cash to buy the external elements of livelihood such as fuel, oil, cooking materials, matches, salt, cotton, cloth, or whatever they may be, and to meet other obligations. Even this type of self-provisioning economy, to be found where the level of mercantility is not high, requires access to a certain minimum land area in order to provide what is locally regarded as an acceptable livelihood, and it seems likely that most cultivators in poorer countries fail to achieve even these modest goals.

An increasingly common situation is that of the land-poor cultivator who cannot support the livelihood of his family out of his own food production plus purchases out of cash income from sales, and must sell labour as well. Indeed, as landholding becomes more restricted, and rents more extortionate, an unpredictable and patently inadequate livelihood is improvised out of varying combinations of the elements mentioned above and supplemented by the economic activities of members of the household, such as craft and artisan work, exchanges and loans, seasonal migration, petty commerce, marginal pasturing of animals, gathering, hunting, fishing, and so on.

Or, there may be a partial dissolution of the family group with the younger members seeking livelihood independently. It is largely the irregularity of livelihood due to the fluctuating quantity of food availability that pushes this category of cultivator into debt and consequently restricts his entrepreneurial scope still further by the imposition of personal dependence, and by the diseconomies of hand-to-mouth living without reserves.

Key to Motivation

The assorted economic means upon which livelihood is balanced also involve the maintenance of a set of social relationships and obligations, so that any radical change must be examined as to its compatibility with the maintenance of these bonds, which may have a very inequitable character.

It is therefore misleading to base calculations about how cultivators are motivated on an image of their persistent profit-seeking, though this can indeed be the dominant motivation of very small cultivators, as was demonstrated in the case of Taiwan (*Wang and Apthorpe, 1974*). It is commonly found in practice that families in locality groups consisting of small cultivators are at any particular moment preoccupied, even obsessed, with some particular set of problems involved in pursuing livelihood that makes straightforward profit-seeking a remote dream.[3]

While the members of a competitive rural society with ample technical and natural resources are likely to manifest

[3] A generic indication of the types of entanglement faced would have to include petty service monopolies, indebtedness, structural subjection and isolation or exclusion from agricultural services and facilities, absence of food reserves causing distress sales and of other resources or profit-depleting emergency measures in bad years or after bad harvests, sickness, damaging price fluctuations (especially in relation to input costs, food prices, and crop prices). Field studies suggest that the poor cultivators are fighting a defensive battle for a minimum livelihood to hold the family together, and that all but a few families are obliged to retreat step by step.

an 'ascending' scale of livelihood models, where the majority of cultivators are fighting a defensive battle, the idea of an adequate minimum appears, as exemplified by the word *cukupan*, reported by Gillian Hart (1977) from Java, which is thought of in Javanese villages as 'an adequate peasant livelihood' that can be supported by a given area of double cropped paddy-land plus a fish pond, in normal times.

Field researchers in peasant societies are usually able to obtain generally accepted ideas of what the minimum land or food basis of livelihood is considered to be, as did Franke (1972), also in Java. Van der Kloet (1975) made a similar calculation in Morocco on the basis of interview and observation, taking into account daily amounts of cereal consumed by family members and considered adequate, plus an amount of grain sold to purchase non-cereal foods and other necessities of life, and an amount retained for next year's seed. In Chile, in dry wheat-growing areas, similar calculations and consultations by the author produced a figure of 20 sacks or 2,000 kg. of wheat as an amount necessary to sustain adequate livelihood of a Chilean family, by consumption and exchange.

The key to the motivation of the cultivator is therefore taken to be his struggle to defend and improve family livelihood according to appropriate local models, and it is within this framework that we shall enquire what are the main influences upon his decisions.

For the typical rural head of family, the central problem of livelihood is one of how to produce or obtain by exchange or purchase sufficient food all the year round to maintain the members of the family in a condition of 'normal' good health. The possibility of doing this depends on:

—the family's success in securing access to sufficient land and water, of getting together the tools of production, the draught power necessary for tillage, and the other necessary physical inputs;

—the health and morale of the labour force of the family, and their ability to adapt both traditional knowledge or husbandry and appropriate scientific knowledge to existing conditions;

—the bargaining power the producer brings to the market, and the terms on which he can buy his requirements both for production and for consumption, and the price he can command both for his labour and for the produce he offers for sale.

Peasant Livelihood and Government Policies

Peasants and rural labourers know a lot about this food problem—most of them spend their lives searching for a solution to it—but it is not the same as the food problem of governments, nor as that defined by agricultural business. Nor is it the same for peasants as it is for commercial farmers, whose interest in low-cost labour is not inhibited by the value they set upon their own services and those of their family group.

Fortunately, there is some coincidence of interest between government, peasants, and labourers, otherwise our labour would be in vain. Where the main responsibility for the production of the nation's food lies squarely on the shoulders of peasant cultivators or small farmers, governments are interested in the efficient production of increasing food surpluses. And though governments throughout history have used extortionate fiscal policies to squeeze the maximum out of peasants, yet in an age when technology offers the prospects of greatly increased land productivity, the satisfaction and rewards enjoyed by the primary producer are the surest guarantee that he will take steps to produce more, so long as these steps do not interfere with some essential component of his livelihood. The profitability as well as the yields obtained by the small producers are inevitably a concern for governments, although such solicitude may not be found within 'bi-modal' agrarian structures,[4] where over-all national food and export requirements can be achieved by the large farmers using capital intensive methods.

Governments may also have a political concern for the livelihood of the rural poor, either as a result of fear of instability and unrest or out of a reliance on peasants and labourers for political support.

---

[4] That is to say, structures with a large-scale, urban-based, well-capitalized sector, probably devoted to export monoculture, and a qualitatively distinct partly self-provisioning *minifundio* sector, whose surpluses go into local and national food circuits. (For a further discussion of bi-modal structures, see Chapter IV.)

However, farmers and rural cultivators are not the exclusive claimants on government favour. It is axiomatic that a government's central criterion for action is to continue to govern. This requires the support or consent of sectors within the polity that exercise decisive strength at the particular moment, and the acquiescence or control of the remainder. So, any fundamental measure or policy must pass the test of whether or not it is consonant with the interests of its actual or potential supporters. If it does so, then it can be expected that it will be translated into action insofar as the supporting sectors are implicated in those social organs that will carry it out. If it is contrary to the interests of these sectors, then even though it finds patches of support and even though conscientious officials attempt to put it into practice, it is not likely to become widely operational.

In order to be successful, programmes of technological improvement or transformation must have political motors.

What we shall find to be most problematic and of greatest weight is the pattern of linkages between those sectors that support the government and the various classes and categories that participate in the productive process—labourers in the field, self-provisioning owner-operators and tenants, land-owners, entrepreneurial farmers, 'service-entrepreneurs', national and transnational agribusiness, and the rapidly expanding bureaucracies themselves.

It is the relative power of these groups in relation to the whole polity, and their links with the government, that is likely to make, metamorphose, subvert or break policies and programmes. This political aspect is illustrated in the next chapter.

The Market and the Cultivator

Most industrialized countries, both centrally planned or market economies, already manifest productive relations that can absorb significant advances in agricultural technology without great stress being put upon social relations. In these countries, the Green Revolution is not news. But in those countries where agrarian relations continue to be largely pre-capitalist, and where the livelihood of the majority of country people is supplied directly by their link with productive

activities involving self-provisioning as well as market agriculture, the new technology initiates—or rather accelerates—
far-reaching structural changes.

The most trenchant factor of change is the origin in the
industrial sector of the essential components of the new
technology and especially of fertilizer. This requires a difficult volte-face for the self-provisioning economies whose
inputs have hitherto originated on their own farms or in their
own neighbourhood. In order to take advantage of science's
contribution to agriculture, the cultivator must operate in the
market to obtain and pay for the new elements of production.

—Can he continue to have assured access to sufficient land
to maintain production for domestic use *and* to become a
market operator?

—Must he give up self-provisioning and operate entirely in
the market?

—If so, has he the reserves to resist shortages, late delivery
of inputs, rising fertilizer prices, and can he pay off the debts
incurred in his productive operations to banks or money-
lenders, and at the same time provide his family with their
daily rice or bread?

—Can a marginal market relationship substitute for marginal access to a decreasing plot of land?

Since all improvements in productive technology must be
performed by the cultivators,[5] in our discussion we place

---

[5] By cultivator is meant the person (man or woman) who performs the role
of organizing and carrying out the sequence of acts that make up agricultural
production. He must ensure access to land and water for the period of the crop,
he must possess a 'production recipe' or body of technical knowledge about the
tasks to be performed, the inputs to be used, the timing of the tasks in relation to
growth rates, weather, soil types etc., and the precautions that must be taken
against the numerous threats to his crop. He must make dispositions and investments in order to obtain seeds, manures, draught animals, tools, and other objects
necessary to the productive process. He is responsible for seeing that the appropriate labour is available, whether it be his own or that of his family and dependents
and neighbours, or that which he can contract for wages. And when the crop has
been harvested, the cultivator will settle outstanding accounts with those whose
goods and services have contributed to the process of production, and retain the
balance. Cultivatorship is a kind of 'natural entrepreneurship'. At its most simple,
the tasks of cultivatorship may all be performed by a single individual. In the case
of most small farms they are performed by the head and members of the household,
with some additional labour on an exchange basis or in return for wages in kind
or cash. In larger farms, labour becomes separated from the other aspects of

them firmly in the centre of the stage and we assume that their decisions are made with a view *to defending and if possible improving family livelihood.*[6]

Another reason for placing the cultivators at the centre is the fact that they constitute the poorest one third of humanity and that their poverty and low productivity are two sides of the same coin. Improvement of their productive capacity can improve their livelihood and vice versa. The central problem of human development is here.

Market Incorporation

In addition to requiring a different and more complicated husbandry, the new technology, on account of its use of commercial inputs, calls for the sale of an increased proportion of the harvest, and arrangements to obtain operating capital for the much higher expenditures required on inputs. Thus the 'adoption of technical innovations' is not all the story.

Those among the cultivators who have hitherto produced mainly for the direct provisioning of their families must also face a brusque change in their mode of operating economically. They must go into business. It is possible that the transformation from self-provisioning to market-oriented agriculture is the greater obstacle to adopting the new technology. They must come to terms with *market incorporation*.

The process of market incorporation may be thought of as representing a passage across a wide frontier or perhaps a transitional belt, between a locality-bound, self-provisioning, pre-capitalist mode of production and a market-motored society. On the one hand, as we have already expressed it, the national market complex in most countries exercises an incorporative pressure on the local communities, families, units of production, customary institutions, groupings; on

---

cultivatorship and large 'business' farms present a picture of sharp and hierarchical division of function, with labour, management, entrepreneurship, and possibly land-ownership all performed by different persons.

[6] This is not at all the same as maximizing profits, which is likely to be the dominant aim of the business farms and which emerges as farms become increasingly market-oriented in a monetized society.

the other, individual cultivators who retain or have acquired a more ample resource base learn quickly how to come to terms with the market society and to make use of its opportunities and facilities.

This does not necessarily make them missionaries of the new commercial order. It is equally likely that they will make use of their own resources and the advantages and controls they enjoy, based on custom and privilege rather than free competition, to enhance their business position, ensure a supply of labour, and corner the maximum possible share of government-granted facilities.

II POLITICAL MOTORS OF TECHNOLOGICAL INNOVATION

The importance of the political will to introduce innovation in food production is demonstrated, first in the framework of three colonial regimes, and then in the pioneering stages of the development of the new varieties of wheat in Mexico.

The Green Revolution as a strategy for development was directed towards tropical Third World countries whose economies had been deeply warped by their colonial experience and whose development paths sought economic reorientation and political autonomy.

During the first half of the present century, scientific research and technological development for tropical agriculture was dominated by the metropolitan powers' quest for increasingly profitable production methods for those exotic products of the tropics for which the metropoli yearned and the colonial subjects sweated—tea, cocoa, cane sugar, pineapples, dates, bananas, spices, rice, cotton, jute, and so on. A whole suburban world of brokers, shippers, underwriters, warehouse clerks, bankers, retailers, and *rentiers* lived prosperously on the differentials between the cost of production in the fields of Asia, Africa, the Caribbean, and South America, and the selling prices in the metropoli. And while local economies in these countries were bent to commercial export production, the production of local crops continued in deteriorating conditions with no benefit from scientific research.

It is illuminating to look at a number of cases in which colonial powers have attempted to introduce the new technology in their own tropical possessions.

I. POLITICAL MOTORS IN COLONIAL SITUATIONS

(i) Taiwan

It was the Japanese who made the most decisive moves in

recent times to improve the rice yields and who were the pioneers of the chemical-genetic technology. In 1895, they occupied Formosa (now Taiwan). The period coincided with the expansion of urban populations in Japan under the influence of the rapid growth of industrial capitalism, and an internal demand for food that Japanese agriculture could not meet.

The Formosan agricultural product of most interest in the circumstances was sugar, its rice not being palatable to the Japanese and hence not marketable. Nevertheless, Japanese administrators appreciated the potentiality of Formosa as a supplier of rice to their country, and undertook a long-term strategy of imposing upon Formosan agriculture the production of a rice for export having the qualities appreciated in Japan.

Japanese sources claim that agricultural research was begun shortly after the acquisition of Formosa. A general survey of all rice varieties planted in the island was undertaken and took ten years to complete. The results given in 1915 showed that there were as many as 1,197 native varieties being cultivated at that time. Technical and administrative efforts to improve the quality and yield of rice began in 1909, and the next year a 'Native Rice Improvement Programme' was initiated. Its first task consisted of eliminating as many of the poor-quality varieties as possible and accustoming the farmer to abandon the promiscuous reproduction of local varieties in favour of selected seed. It also meant the eradication of red rice, a sturdy drought-resistant strain, but unacceptable to the Japanese consumer.

The first step was to select a productive variety of the old native rice that was stable, less chalky, and easier for drying than the typical Formosan variety, and which had the soft quality preferred by the Japanese. At the same time, a harsh campaign against red rice and other varieties was conducted by the police. By 1920, the number of native varieties had been reduced to 390.

In addition to the development of new varieties suitable for the Japanese market and therefore commanding a better price, irrigation was improved and extended. Deep ploughing, intensive planting, and the use of artificial fertilizers, green manure, and pest measures were propagated by the Japanese. However, erratic climatic conditions frequently damaged crops and the overall increase in total gross farm output between 1906 and 1920 was only 1.7 per cent per annum compared with a population growth trend of 1.2 per cent. In the fifteen-year period, the average yield of rice per hectare for the whole island increased by no more than 6 per cent.

Although many of the necessary conditions were established for important advances in technology and productivity as regards scientific research, infrastructure, marketing, and the organization of production at the local level, no jump in land productivity had occurred. However, in 1922 the plant breeders were successful in producing a cross between *chailai* and *japonica* varieties that combined the three main traits required, namely marketability, high fertilizer responsiveness, and quick maturation. This new variety, which was given the name of *ponlai*, took 13 days to mature in the first cropping and 102 in the second, in each case about 30 days less than the current varieties. This opened up possibilities for third and fourth crops within the year, and thus afforded a land-saving quality that was to become more important as the increasing rate of population growth pressed upon scarce resources in irrigated land.

In fruition of efforts invested between 1906 and 1920, the period that followed (1920–38) was marked by exceptional advances. *Ponlai*, which covered 400 hectares in 1922, was planted in 68,700 hectares in 1925, 131,200 hectares in 1930, 295,800 hectares in 1935 and 324,000 hectares in 1940, by which time it covered more than half the island's total rice area. The use of artificial fertilizer increased by 254 per cent, the average rice yield increased by almost 50 per cent and an average increase of 4.3 per cent per annum was achieved in gross farm output, in which improvement in the yields of sugar also played an important part.

The development of rice technology in Formosa by the Japanese colonial rulers and the Taiwanese peasants and technicians had a very important influence on the spread of improved technology in Asia, once World War II was over and reconstruction could proceed. As far as Taiwan's own post-war evolution is concerned, an ample picture is given in the relevant UNRISD study, which was independently presented in Taipei. (*Wang and Apthorpe, 1974.*)

(ii) Northern Rhodesia

Another colonial initiative described by a contributor to Global Two (*Geza, 1975*) was the campaign for the adoption of hybrid maize with a high yield by expatriate commercial farmers in Northern Rhodesia. Here the drive to introduce the new technology and to raise production was generated by the demand of the Copperbelt mining companies for an assured supply of maize at a reasonably stable price for their African miners. This became a matter of special importance on account of the uncertainty of the rainfall in the maize-producing areas of the country.

Data supplied by the colonial Department of Agriculture show that average yields between the years 1947 and 1955 varied between 990 and 1,485 kg. of maize per hectare. The main suppliers of the miners' food needs were the expatriate commercial farmers who occupied the flat lands on both sides of the railway running through the Central and Southern Provinces. Their yields, however, were not only fluctuating but also on the average very low, and from 1951 onwards they were unable to produce enough maize to feed the urban consumers. Grain was shipped in from the Eastern Province in spite of the high transport costs involved. Although this relieved the shortage, consumer prices rose by 265 per cent during the eight years between 1945–6 and 1952–3.

A typical consequence of the high price of maize was a series of labour disorders in the mines, causing a decline in copper production and consequently a fall in the profits of the copper-producing companies. Thus, the poor performance of maize-producing technology damaged the most significant interests of the colonial power.

The Department of Agriculture identified a widely used open-pollinated maize, Hickory King, as the main obstacle to an improvement of yield in the commercial sector, and took steps to find an improved variety. Between 1950 and 1954 experiments were carried out with the Southern Rhodesian Double Hybrid variety, which showed improvements in yields between 2 and 54 per cent.

The new variety was distributed to farmers in the commercial sector, though the peasants continued to use traditional
seeds. On the basis of incomplete data, it appears that yields
trebled between 1954 and 1964, while those in the peasant
sector recorded no advance.[1]

The post-independence government adopted a policy of
encouraging the widespread adoption of high-yielding varieties
of maize by peasant cultivators. Geza cites two factors seem-
ingly connected with the end of colonial rule that seem to
have influenced this policy: a large-scale exodus of colonial
farmers as independence approached and a sharp rise in
migration to urban areas, increasing the demand for food in
those areas. Both of these circumstances increased the
country's dependence on the peasant sector for its food, and
promoting new varieties was an obvious starting point for
increasing yields in the sector.

(iii) Surinam

The third example of political motors operating in a colonial
situation comes from Dutch Guiana, now Surinam. In the
development of farming in Europe after World War II, optimum size of holding increased, creating a demand for land
and farming possibilities amongst farmers' sons. The Dutch
Government therefore sent a committee to the sparsely
populated Dutch Guiana to study the possibilities of settling
Dutch farmers on the Coastal Plains. The Committee concluded that a Dutch farmer could make an adequate livelihood
farming mechanically an area of 100 hectares mainly in rice,
but with supplementary cacao and coconut trees.

In 1949, an experimental *polder* (an irrigated and drained
level cultivation area) was created that ultimately covered an
area of 500 hectares, and a rice breeding station with an
estate annexed to it was established on the *polder*. In the
same year, a Society for the Development of Mechanical
Farming in Surinam was founded to organize a programme of
settlement by Dutch farmers. Further *polders* were built.

---

[1] Our researcher in Zambia cautions that many other improvements in farming
methods—new planting procedures, optimal weeding, use of insecticides and fertilizers—must have played a role in whatever yield increases had occurred (*Geza,
1975*, p. 79).

However, it was found that rainfall was too heavy to allow the cultivation of dry crops, so that prospects were limited to the monoculture of rice, which could be cropped only once in the year.

Meanwhile, the relationship between metropolis and colony was changing and in 1954 a constitution was introduced vesting Surinam with qualified internal political autonomy. The plan to introduce Dutch farmers was scrapped, and the main Wageningen estate, consisting of approximately 6,000 hectares, became a single mechanized production unit managed by the Society; it went into full production in 1956.

The basis for the use of improved seeds was laid by a plant breeder named van Dijk, who developed new varieties between 1941 and 1946. His materials were used and crossed with imported varieties during the 1950s and, by the end of the decade, two suitable varieties had been produced that were nitrogen-responsive, had stiff stems suitable for standing up to combine harvesters, and a long grain suitable for marketing in Europe. However, it was only at the end of the 1960s that it was possible to produce quick-maturing varieties and so make double-cropping possible. The *acormi*, *awini*, and *apani* varieties were available in 1970, matured in 100–120 days, had a short stem, and were more nitrogen-responsive than the earlier varieties. These varieties reached yields of 4–5 tons per hectare in the farmers' fields during 1972, which was a good year.

Wageningen Estate is the world's largest mechanized rice-producing farm, and operates as a non-profit-making concern. It is also a community whose services are financed by gains made by the enterprise, which, however, has been in need of subventions from the Dutch Government for every year of its operation (up to 1970) with the exception of 1968.

In 1971, a policy of 'Surinamization' was adopted, and the higher echelon managerial and administrative staff from Holland were replaced by Surinamese.

Wageningen is an extraordinary anomaly, owing its existence to the discontinuity of history and to the decline of colonialism.

In each of the colonial cases described, the motivation stands out clearly: the programme is planned to benefit an important sector of the metropolitan society, the economic and social impact on the society in which the initiative took place being incidental, *yet the outcome in each case was quite different from the aim envisaged by the programme's promoters.*

In Surinam, the prospect of decolonization and the rate of European recovery put Surinamese, not Dutch, cultivators into the 30-hectare irrigated farms, where they rapidly became rich beyond all expectations. In Zambia, although many of the expatriate commercial farmers left with decolonization, their places were taken by better-off or politically influential Zambians, and the bi-polar dichotomization of production between large capitalized farms and a poor self-provisioning sector with communal tenure system has dragged on: this inevitably weakened peasant farming since the commercial farms were more capable of providing food for mines and other urban dwellers and were better prospects for credit and for technical assistance.

In Formosa, the Japanese colonial masters were replaced first by the Americans and later by Chiang Kai Chek and his mainland followers. Agricultural production, at the centre of development planning, was firmly assigned to small-scale commercial cultivators, and no land magnates were able to control dominating expanses of land or, as a class, to monopolize credit and services.

Taiwan's claim to have achieved broadly-based rural development is eagerly debated, as are indeed the explanations of how this state of affairs has been achieved within the private enterprise system.

The UNRISD studies draw attention to several aspects of the situation that have hitherto been overlooked: none of the three dominant external élites (Japanese, Americans, and Mainland Chinese) have held a brief for any existing or emerging group of land proprietors or large-scale farm operators and, as a result, in order to build up production, the administration has taken seriously the equity-based agricultural and infrastructural services that it has controlled or regulated.

This is in contrast with those countries with a hierarchical peasant structure or a landlord–tenant structure, in which the scene is dominated by the local magnates able to capture the lion's share of government assistance and to subvert to their own use the measures and facilities undertaken to benefit the common cultivator. It is also in contrast with countries that have a 'bi-modal' structure.[2]

For several generations, Taiwanese farmers have been export-oriented entrepreneurs as a result, it is claimed, of the situation of their ancestors, who colonized the island in order to supply the mainland with commercial crops. The implication is that they had already made the economic leap away from self-provisioning, and were able to concentrate on their market operations without the difficult choices that tie up the self-provisioning peasant.

The third important fact, which will be discussed briefly later, has to do with the existence of an 'entrepreneurial complex' consisting of infrastructural and production facilities and services controlled by the government, with participation by farmers' associations.

II. POLITICAL MOTORS: THE MEXICAN CASE

The interaction between political and economic forces and technology, which traces its zigzag path through the story of the development of Mexican agriculture since the beginning of this century, is a fascinating and revealing story, well told and interpreted by Cynthia Hewitt de Alcántara (1976). The most spectacular part of the story is the evolution of cereal production in the dry North-Western provinces of Sonora and Sinaloa. Between the five-year period 1925–30 and that of 1965–9, national average yields for wheat rose from 686 to 2,415 kg. per hectare. During this period, the area under wheat also increased by approximately 50 per cent. As a result, Mexico began to export wheat in 1955. This unusual feat is one of the sensations of what was later called the Green Revolution, and its most spectacular achievement took place in Sonoroa and Sinaloa.

---

[2] Such countries are to be found especially in Latin America, North America, and the Caribbean. Effective command over resources exercised by the commercial sector, combined with its political power, tends to strangle the flow of resources to the small producers, whose economies are undermined and whose communities are marginalized.

The first important consideration is the timing of the technological revolution. From 1910 to 1940 Mexico passed through a period of revolutionary change, at times violent, at times fitful and patchy, which in large part destroyed the neo-colonial society of manorial estates (*haciendas*) and debt servitude (*peonaje*). Under the presidency of Cárdenas (1934–40), extensive measures were taken to institutionalize a mixed agrarian order based on collective landholding, with both collective and family entrepreneurship and private property subject to limitations of size. The year 1940 marks the election of a president representing opposed trends and the beginning of a period of boom conditions for raw materials, as a result of World War II economic requirements and scarcities.

The discussions that took place in 1941 between Mr Daniels, US Ambassador to the Government of Mexico, and Mr Henry Wallace, Vice President of the United States, and which set in motion the scientific mission to assist in the development of agricultural technology, also ushered in an era of good relations between the US and Mexico following a period of open and noisy hostilities.

As a consequence of their negotiations, the Office of Special Studies was initiated in Mexico in 1943 as a co-operative venture between the Rockefeller Foundation and the Mexican Government, and a plant-breeding programme set in motion to increase the yield of corn and wheat and to provide a variety of wheat seed that would be resistant to the rust-disease. This joint research enterprise, and those arising out of its work, were to be responsible for a steady outpouring of important technological innovation throughout our period.

However, an equally important component of this evolutionary process had already got under way during the Cárdenas regime—state investment in irrigation and support through credits for private irrigation works and machinery. From 1936 to 1973, the irrigation of farmland spread from about one million to 4.7 million hectares, the main difference between the pre- and post-1940 policy being that, before the crucial date, irrigation schemes financed by the Government had been directed to benefit *ejidos*, or collective landholdings, while after that date irrigated land was sold off to a mixed assortment of middle strata buyers, which included both speculators and entrepreneurs.

The steady process of irrigation in the North-West seems to have induced conditions of a gold rush since it 'created' natural resources year by year that generated waves of speculation and land dealing. However, with the yearly appearance of new high-yielding seeds, even the so-called 'nylon farmer' was likely to do well in his newly-irrigated land though rust-disease was still a threat. A securely rust-resistant strain was released only in 1954.

The dynamic of growth in the North-West was pushed along by the boom conditions created by a heavy programme of government spending on irrigation in the area, which reached its peak from 1953 into the early 1960s. To complete the 'package', a felt need for fertilizer was generated as controlled irrigation replaced flooding, since without flood alluvia, plant nutrition diminished. But the capital investment involved in the use of purchased fertilizer was at risk until the rust-resistant strains were secured. However, success with the rust-resistant seeds given the optimal dosage of fertilizer was muted by the fact that the ears became too heavy for the structure of the plant, which tended to collapse under the strain. Only with the production and distribution in 1954 of a dwarf rust-resistant plant, to be used with controlled irrigation, were important yield increases achieved. Nevertheless, by that year, 90 per cent of all wheat lands in the North-West States of Sonora and Sinaloa were sown in one of the new 'man-made' varieties, while *Lerma Rojo* and others had replaced these on 88 per cent of the expanding wheat area by 1960. Thus a dynamic process had been set in motion (*Hewitt de Alcántara, 1976*).

Until the late 1940s, Sonoran wheat suffered disadvantages in the market, firstly because of its low milling quality, which held down its prices to 70 per cent that of imported wheat, and, secondly, because of the transport costs of the long haul to the mills in Central Mexico. But by the end of the decade, a uniform guaranteed price was offered for all kinds of wheat, and transport to the mills from the North-Western States was subsidized. In 1952, President Alemán went further and established machinery empowered to assure purchase of all wheat produced. (This type of guarantee has become a permanent feature of the wheat market.) However, the subsidy element in the guaranteed price was lowered in 1965 and a substantial fall in the area devoted to wheat took place.

In addition to very heavy government investment in irrigation, which reached its maximum intensity between 1949 and 1958, both transport and electrification were heavily financed. The railway system was overhauled and rehabilitated, and a network of farm roads was built for which the government paid two-thirds of the cost. This raised the demand for petrol, the annual consumption of which jumped from 1.5 million litres in 1946 to 5 million litres in 1948, making necessary special concessions by the government permitting direct importation from the United States. Use of electricity also doubled between 1950 and 1955.

In an attempt to explain the great Mexican wheat revolution, let us now look at the technological and political inputs side by side, at the evolution of the entrepreneurial class, and finally at the nation.

The great Mexican wheat revolution hinges on the twin motors of profitability and political support for the entrepreneur group from the government.

The scientists produced three aces:

(i)   a series of varieties that were increasingly responsive to fertilizer, and hence raised the profitability of additional investment;

(ii) a lowering of risk by building rust resistance into their latest varieties;

(iii) elimination of the risk that heavily fertilized strains might lodge, or collapse, under the added weight of grain.

The government, on the other hand, in fulfilment of its post-1940 free enterprise policy[3] and its commitment to the class to which the new entrepreneurs of the North-West belonged:

---

[3] President Manuel Avila Camacho (1940–6) therefore hastened to assure the nation, in his inaugural address, that he would base the future of Mexico 'principally on the vital energy of private initiative' and that he would most especially 'increase protection of private agricultural properties, not only in order to defend those already in existence, but also to form new private properties in vast

—invested in the great expansion of irrigation;
· —established guaranteed prices for wheat;
—built up transport networks from the irrigated region to the market far to the South, and financed the spread of electric power;
—provided sources of credit for farming operations and for private irrigation schemes;
— facilitated supplies of fertilizer;
—fostered mechanization by special exchange rates.

In fact, as land ownership became so profitable in the irrigated North-West under the new wheat technology, intense competition for land and inputs ensued, driving prices up and bringing about the concentration of property. No exact statistics on ownership of irrigated land in the region are available and, after reviewing the evidence, Hewitt suggests that the number of serious operators has been reduced to less than 200, averaging some 500 hectares each.

Thus the great wheat boom that made Mexico into a wheat-exporting country forced the smaller individual cultivators out of business and concentrated capital-intensive agriculture into the hands of less than 200 'millionaire' entrepreneurs.

The career and fate of *ejido* agriculture under the blast of the Green Revolution is discussed in Chapter IV.

III. THE SPREAD OF THE HIGH-YIELDING VARIETIES

The planting on a commercial basis of seeds from the two international centres—the International Maize and Wheat Improvement Center (CIMMYT) in Mexico and the International Rice Research Institute (IRRI) in the Philippines—first took place in the crop year 1965–6. Nine years later these seeds already covered important proportions of the area planted to wheat or rice in many of those countries in which they were of prime importance for food, as is indicated more precisely in Table 3.

In this table, Dalrymple classifies as HYVs (i) those varieties

uncultivated areas.' Agriculture was to play a new role, not as the basis for rural development, but as the 'foundation of industrial greatness'. (*Hewitt de Alcántara, 1976*, p. 19.)

TABLE 3

Percentage of Rice and Wheat Croplands
under HYV, 1972/3

	Rice	Wheat		Rice	Wheat
Afghanistan		18.4	Malaysia	38.0	
Algeria		27.9	Morocco		13.4*
Bangladesh	11.1	17.7	Nepal	14.8	65.8
Burma	4.2		Pakistan	43.4	55.9
India	24.7	51.5	Philippines	56.3*	
Indonesia	18.0		Sri Lanka	2.5	
Iran		6.9	Syria		21.2
Iraq		22.9	Thailand	4.9*	
Jordan		10.0	Turkey		8.0**
Korea (South)	15.6		Tunisia		10.4
Laos	5.5		Vietnam (South)	32.1	
Lebanon		31.3			

*Based on unofficial estimates
**1971/2 figure
Source: Dalrymple, 1974, I.

developed during and after the mid-1960s; (ii) descendants, via national development programmes, of varieties distributed by IRRI and CIMMYT; (iii) previously developed varieties distributed by IRRI and CIMMYT (e.g. Taichung Native I rice from Taiwan province provided to India in 1964 by IRRI); and (iv) some Italian wheat varieties with some of the same characteristics as Mexican varieties. Since the present study uses much Dalrymple data, it follows the Dalrymple definitions, except where specially noted.

In 1972/3, excluding the countries with centrally-planned economies, Mexico and Taiwan, there was a Third World total of 16.8 million hectares of HYV wheat and 15.7 million hectares of HYV rice. Almost all the area under HYV seeds (about 94 per cent of wheat and 98 per cent of rice) was in Asia, of which nearly half was in India. Within Asia, excluding the countries with centrally-planned economies, HYVs occupied nearly 35 per cent of the total area sown to wheat and 19.5 per cent of that sown to rice (*Dalrymple, 1974*, I). In Mexico in 1970 there were 762,500 hectares under wheat. In the preceding twenty years, wheat yields had tripled as different elements were introduced and adopted. Very little of the total wheat area remained unaffected.

In view of the criteria used for defining HYVs for the tables, and of the growing complexity of patterns of heredity as ever new strains develop from crosses between local varieties and HYVs, no attempt is made to present statistical data on plantings subsequent to the 1972/3 season, since the distinctions cease to signify. It is now common practice for governments to turn to the international research centres or to their own national centres for new varieties that can be expected to give improved yields and to embody combinations of desired characteristics fitting specific environments.

Most of the new varieties that constitute the new technology in wheat received their major development at CIMMYT in Mexico and those in rice at IRRI in the Philippines.

CIMMYT was set up in 1966 on the basis of the joint Rockefeller Foundation and Mexican Government programme first established in 1943. The original programme had as its goal the rapid development of a thriving commercial wheat sector within Mexico. The new organization leaves the internal application of HYV research to Mexican institutions and addresses itself to its new global responsibilities for research, development, and training in wheat and maize technology.

The exportable wheat seeds were the result of nine years of experimenting with Japanese *Norin* wheat. Attention was first drawn to semi-dwarf wheats growing in Japan in 1873. In the early twentieth century, some found their way to Italy, where they were bred into a number of improved varieties, later widely used. Meanwhile, the Japanese themselves crossed these varieties with several American seeds, from which a *Norin* variety was released in 1935. It was brought to the US in 1946, crossed with American seeds, and taken to Mexico in the early 1950s, where it was used by Rockefeller Foundation staff to develop the well-known Mexican varieties. New short-stemmed varieties, high-yielding, non-lodging, and more rust resistant, first greatly increased production on Mexican fields and eventually served as the genetic base for the new technology in wheat in many other countries.

From the mid-1960s, as word spread of the Mexican successes, there was worldwide interest in increasing food production by improved technology. The new HYVs, in combination with other agricultural improvements, began to be regarded by many nations as an instrument to raise yields. The CIMMYT programme led to international interest in a similar project for rice. The International Rice Research Institute was established at Los Baños in the Philippines in 1960, financed by both Ford and Rockefeller Foundations and a number of Asian governments, with the active participation of the government of the Philippines.

The Institute was able to draw on a great number of improved rice seeds developed over the centuries, notably in China and Japan. Breeding programmes at IRRI have produced several series of HYV seeds that are widely used in rice-growing countries, though locally improved, non-IRRI rice seeds have also been responsible for good increases in production: H-4 in Sri Lanka, ADT-27 in India, Agriculture 5 and 8 in North Vietnam and other important varieties in Egypt, Brazil, Surinam, and China, for example.

The fact that it has been possible to keep systematic records of the spreading use of the seeds released by the international centres must not be allowed to give the impression that the work of many of the national centres is without importance.

PART II

III COMMUNAL TENURE STRUCTURES AND AN AFRICAN EXPERIMENT

Some background is provided to changing agrarian struc-
tures in Africa, and the introduction of new rice varieties
in Sierra Leone is described.

I. THE NEW RICE IN AFRICA

Government measures to introduce the new technology and
the response of cultivators to these measures can be understood only if they are examined within the dynamics of
differing socio-economic and agrarian contexts. We have
already proposed three major varieties of agrarian structure as
appropriate for this analysis. Material on the introduction of
the new technology in a number of countries will be discussed
and, as far as possible, evaluated within this framework.

The first major category of agrarian structures includes
those that are emerging directly from traditional communal
tenure systems. These systems still predominate in Africa
though they are also to be found in other continents, usually
in areas least penetrated by market forces. In view of the continuing importance of these structures and the special problems involving technological development in them, some
background on the African situation follows.

Most social systems in pre-colonial Africa exhibited two
related features in connection with the production and
accumulation of wealth. In the first place, a person's rights
to land and its produce were part of his birthright. His
ascribed rank within the local community or lineage
structure determined the exact extent of these rights. But at

the same time a person could manipulate his rank by various mechanisms—enhancing his prestige through particular skilled exploits such as hunting and warfare, acquiring more dependants by organizing successful marriage alliances or by being a skilful politician. The more powerful a man was in terms of dependants the more the goods that were produced 'for him' but, conversely, the more he had to redistribute in order to maintain his legitimacy and his position.

There were thus two sets of operative relationships determining individual decisions about production and exchange: rank and reciprocity. Rank was important because only through it could wealth be acquired. Meillassoux (1970) writes of the Gouro:

According to Gouro traditionalists, wealth is a social accomplishment. . . . The association of wealth with rank follows logically from the relations of production which concentrate in the hands of the oldest member of the community the products of his dependants' labour. It comes then, not from the elder's own work, but from the importance of the group he controls. . . . But if wealth cannot be acquired independently of rank, the latter cannot be preserved without wealth.

Lucy Mair (1957) has written in similar terms of the Ganda, amongst whom chieftainship was the most important means of acquiring wealth. This possibility was, however, more commonly counteracted by the importance of reciprocity in relations both of production and consumption. The low level of technology and agricultural productivity encouraged these reciprocities. The means of completing labour-demanding agricultural tasks were exchanges of labour between families that were usually followed by beer, not implying a payment but rather in the sense of a prestation. Beer was not a substitute for reciprocity in kind. Such reciprocal relations can really be maintained only where there is a situation of relative equality (*Barth, 1970*).

Similarly, reciprocity existed in the sphere of consumption and was a major mechanism preventing both the maintenance of significantly superior living standards by those deemed wealthy, and also preventing the inheritance of wealth by their heirs. Both rank and wealth were derived from the scale of one's dependent group—but in return one was obliged to assist dependants in situations of need.

Colonial Heritage

The growth of colonialism in Africa meant the penetration of African societies by the market economy in various forms and combinations of factors. It was penetrated by the appearance of opportunities for the sale of labour in enterprises sponsored and stimulated by the colonial power or by colonizing entrepreneurs—often accompanied by pressures upon the local economies making it necessary to seek cash earnings; it was penetrated by the demand for cash crops and animal products and the establishment of a marketing infrastructure facilitating their purchases; and, likewise, it was penetrated by the appropriation of lands and resources by intensive colonial producers.

This led to a much greater emphasis on the individual producer, his specific tasks in the production process, and his individualized reward for it. It has also led to a profound disjunction between the persistence of the tribal social system and the system of production on which it was based. The development of a market for primary products outside the tribal system separated the acquisition of wealth by production from rank and weakened the tribe as a system of rights and obligations, and as a matrix within which livelihoods could be achieved by all.

Colonialism has resulted in very confused, insecure, and heterogeneous patterns of land tenure in most African countries. On a background of customary systems in which the amount of land controlled by a single individual varied according to the number of his dependants and his political position, a rapid process of individual appropriation is running its course in all but a few countries. This process was initiated as farmers began wanting to plant cash crops and became able to increase the amount of land cultivated by the adoption of more advanced technology, and by the use of animal and automotive power. Rural populations were also increasing and the planting of permanent crops created a demand for the permanent possession of land and its purchase and sale to provide a secure basis for individual investment in land improvement.

Access to more productive land thus came to be conditioned

by the individual's capacity to pay for it, regardless of social ties and positions. So the control of productively valuable land is acquired by individual property holders and, as land rights derive increasingly from prior accumulation and exchange, a significant differentiation between individual cultivators takes place on the basis of the amount of land individually held by each. These two traits contribute to the declining authority of traditional chiefdom and its customs and the emergence of a new stratum of power-holders based on the extent of land in individual possession.

Penetration by the market and its opportunities has also generated productive commercial enterprises on a larger scale and the corresponding emergence of African agricultural and service entrepreneurs. This movement has almost always been coupled with the individualization of the productive unit, and, frequently, the withdrawal of the individual entrepreneur from the socio-economic liens of his tribal ascription but also from most of the social life that it involved and the authority it imposed.

The individualization of entrepreneurship should not necessarily be regarded as a *natural* process, it being firmly encouraged by the development policies of 'late colonialism' (see *Feldman and Lawrence, 1975*, pp. 170–2) and by the impress of the marketing institutions themselves, designed for dealing with the individualized entrepreneur.

II. THE SIERRA LEONE SWAMP-RICE SCHEME

The problems involved in programmes to advance technology in the kind of conditions described are illustrated by an extensive survey carried out in rural Sierra Leone to analyse the performance of a government programme to introduce new high-yielding varieties of rice.

For generations rice has been cultivated in Sierra Leone on the gentle rolling hillsides that cover most of the country and which enjoy a rainfall of 100 inches annually. In addition to this predominant upland rice cultivation, wetland rice is also grown by traditional methods in swamps and riverain strips. The programmes in question set out to introduce the new technology for the cultivation of rice in the swamps.

The mode of cultivation of upland rice is described as follows: each little community has the characteristics of an autonomous land-group, self-provisioning to a high degree. In his countryside survey, Weintraub (1973) shows that only 20 per cent of all cultivators reported sales of rice, while 20.4 per cent and 17.3 per cent reported sales of palm-kernels and cocoa respectively. This must mean that the bulk of food produced was consumed within the family of the producer or exchanged within the local community. No fertilizer was used and seeds came out of last year's crop. However, to secure a proper land allocation and labour supplies, 'a variety of personal village-level relationships had to be maintained'. (*Weintraub, 1973*, pp. 36–7.)

All persons born to the local lineage have inalienable rights to the amount of land considered necessary for livelihood, whether on an individual or family basis. The head of the lineage is responsible for designating specific areas for the use of heads of families, taking into account the area needed for the exercise of shifting cultivation in that place; in their turn, family heads will allot the usufruct of lands suitable for farms to sons, younger brothers, etc. In case of disagreement, the 'Town Chief' is expected to adjudicate. Should any other person living in the land-group without birthright require land for maintaining his family, usufruct may be obtained from a member of the group in return for a pledge in money or in some cases by paying a cash rent to the local authority.

The foregoing applies to uplands, in which shifting agriculture is practised. Since swamp and riverain lands can be cultivated yearly without being required to lie fallow, and because the supply of swamplands accessible to human habitation is limited, there is a marked tendency for these lands to become the *de facto* property of the families that obtain possession of them. The growth of the custom of pledging land seems to be an accompaniment of this solidifying of ownership, independent of use.

Weintraub claims that where a land-group relies entirely upon swampland for livelihood, relative equality of holdings is likely to prevail. But in numerous localities studied, the main burden of production for livelihood was borne by the uplands while the swamplands available served a supplementary purpose. Rights to cultivate swamplands did not (or had ceased to) form a part of the birthright and 'a steady process of individualization of rights was taking place in those areas which lay outside the operation of the shifting cultivation system serving the livelihood needs of the rural population or which appeared to promise substantially higher levels of production'.

The Inland Valley Swamp-Rice Scheme became operative on a national scale for the 1970-1 rice production season. The government made arrangements to provide seed rice and fertilizer in sufficient quantities to cover up to 8,000 acres and at no cost for the first season; some technical assistance for land preparation, water management, and cultivation practices; and a subsidy in cash towards the cost of clearing swampland.

Tools at the disposal of the vast majority of cultivators consisted of small hoes, knives, cutlasses, and shovels. The programme was launched on a national scale by letters and visits to chiefs to explain what the scheme had to offer to the individual cultivator, and what he and the villagers could expect to get out of it.

The prospective participants had to be assured of access to the necessary swampland and of the labour needed for its initial clearing and preparation. Having earmarked his land, the cultivator next had to get the local extension agent to inspect it. If the land was found to be satisfactory and if the cultivator could convince the agent that he had access to it guaranteed for the next few years and could also command the necessary labour, he was officially enrolled in the scheme.

From the accounts given by the researcher, the initial response to the scheme was encouraging and some 2,500 cultivators applied to participate in the scheme with 6,000 acres. The research survey covered a random sample consisting of 200 participants with an additional 48 non-participants as control, and a rough indication as to whether the scheme was successful is given by the yield figures for the first season referred to below.

The main problem confronting the cultivator was that of obtaining sufficient labour to carry through the land improvement operations and the various practices of the new husbandry. In describing the preceding pilot project known as Dicor, Weintraub (1973, p. 78) says:

The *labour* that would have to be used would be *significantly greater* than in either the upland or the traditional swamp-farming systems. The clearing, felling and stumping process would be more difficult and time consuming in that (a) it would be more complete than was usual in the other systems and (b) it would be undertaken while the farm was in varying depths of mud and water. Second, the land would have to be hoed, harrowed, puddled and generally given a more thorough preparation process than was customarily used in the other systems. Third, since the cleared debris would be too wet to burn, it would have to be pulled off the farm by hand, a task not previously undertaken. The farmer would also have to prepare the nursery sometimes beforehand.

The special difficulty presented by the labours outlined was that they clashed with the traditional upland rice harvest labour needs. Swamp-farming was seen as a supplementary exercise, while the cultivation of upland rice was considered a very central activity upon which the livelihood of all rested. So as the upland harvest was still occupying village labour at the time when land clearing had to begin, the participant in the swampland scheme either had to delay operations or to hire labour from outside, unless he had sufficient political power to coerce the labour he needed. The need for hiring labour was foreseen in the planning programme and a special allowance was to be given to the cultivator as a contribution to these costs.

Figures show that 43 per cent of the participants improved their yield in comparison with the average upland rice yield, 17 per cent equalled it and 40 per cent obtained lower yields. There is reason to believe that yields were underestimated, and it was also noted that rice cooked for the labour force was not included in the calculation of the product. In spite of the fact that in a certain number of cases high yields were achieved, in general the picture was not a sufficiently encouraging one.

The author of the report, relying on his systematic interviews with cultivators, extension agents, chiefs, and other expert informants, discusses the more probable reasons for the relative failure of the programme to achieve its most important target of improved yields.

The most important and most concrete constraint upon performance of the initial pioneer work of clearing and preparing swampland for rice cultivation is certainly not access to the land itself but an insufficiency of labour. This was, in part, foreseen by the administrators of the programme, who offered a cash subsidy to participants, the major part of which was intended for wage payments. This particular constraint became apparent in comparing attributes of participants and non-participants in the scheme: it was found that participants had significantly larger numbers of children over five in their families, and that among the participants a larger number held ranks of authority in the communities. Both these characteristics put at the cultivator's disposal a larger labour force, consisting in the one case of his own children and, in the other, of neighbours and dependants who on account of the cultivator's rank were predisposed by traditional obligation to offer him their labour.

The other two straws in the wind of apparent significance were: twice as many of the participants (compared with the non-participants) sent their children to school (*Weintraub, 1973*, p. 180), although the parental generation of both groups was equally unschooled. This may reflect the higher level of rank of the participants, but it is difficult to separate it altogether from a more positive attitude to the new institutions and the new culture. There is also some evidence about economic activities: more participants than non-participants sold surplus produce on the market, and sold larger amounts —their incorporation into the market economy had gone further; and finally, data on the employment of family and non-family labour showed a clear distinction between small cutivators (size judged by the amount of labourers engaged), who depend mainly on family labour, and large cultivators who depend almost exclusively on wage labour, at higher rates than those paid to workers on the smaller farms.

These data point to the emergence of a commercially oriented larger farmer whose initial advantage is not prior access to land but rather special advantages in being able to assemble a labour force as a result of political rank in the tribal structure. His political position also gives him decisive influence, if he cares to use it, with the representatives of the government administration.

Food Security and Conviviality

The case of the swamp-rice scheme in Sierra Leone is different from most of those described and discussed in this report. In the area chosen, land producing an excellent quality of rice for family consumption is available to rural families by virtue of their accepted membership of a particular kin group, and by arrangement with the titular head of that group. This arrangement is possible on account of the low density of population to land. The cultivator's problem is one of organizing family labour and exchange labour, and in order to achieve satisfactory arrangements for the work of his own holding he must work willingly for the collective, for neighbours, and for kin, so that they in their turn will later work for him.

It is unwise to consider the social conviviality involved in mutual aid simply as means to the economic ends of production since the ensemble of collective labours, land allocations, group celebrations, plantings, harvestings, and thanksgivings form part of what we have called livelihood.

According to Weintraub, a number of cultivators who had already lived and worked in more expansive social situations emerged and were able to make rapid progress as new-style cultivators, motivated to use all their skills to achieve monetary gains with which a different livelihood might be purchased. But the common cultivator, pursuing food-security and a certain quality of conviviality, was not enchanted with what the new technology offered him—on the positive side, some cash for paying labour and a chance of larger rice harvests and, on the negative side, a new technological dependence, some disorganization of his normal agricultural activities in growing hill rice, and also increased contact with government officials likely to extort 'gratifications' from him.

The study of the Inland Valley Swamp-Rice Scheme of Sierra Leone introduces a very important issue for African agriculture, namely whether it is possible for governments to apply scientific principles and industrially produced inputs to cultivators farming within a regime of traditional communal tenure. The experience of the scheme, however, does not provide an answer.

The same problem is touched upon in a brief report by René Dumont (1971) on Mali, in which the reader learns that in the development of the Office de Niger rice-producing lands, the government dealt with communal landholding units rather than with individual family units of production, but no more details of procedures and outcomes are given. In most countries where studies were made, in fact, some form of communal or feudal tenure had prevailed at an earlier stage—and had been displaced by the penetration of private ownership under the impact of European colonization. In the Spanish colonies, the process continued through the seventeenth, eighteenth, and nineteenth centuries.

In Asia and Africa, the process is a more recent one. For Morocco, it is well described by Van der Kloet (1975). And in the companion volume to this, an account of the introduction of hybrid maize in Zambia (formerly Northern Rhodesia) is given.

Under the colonial regime, some of the best lands served by what is known locally as the 'line of rail' were appropriated for expatriate exploitation as commercial farms supplying maize to the mining population of the Copper Belt. No decisive redefinition was carried out after independence, but large numbers of cultivators from the overcrowded 'reserves', which received cultivators displaced by white settlers, have drifted back to become squatters on these lands, in which many of the commercial farms have been taken over by a new class of Zambian farmers. For the moment, then, it appears that the government has opted for the continuance of a bi-modal agrarian structure. The danger of such a policy is, of course, that if the commercial sector appears capable of supplying urban food needs, there may be reluctance to invest development capital on a serious scale in peasant agriculture.

IV   THE DYNAMICS OF BI-MODAL STRUCTURES

*The exploitative nature of bi-modal agrarian structures
is revealed; and the effects of the new technology in
Mexico, Tunisia and the Philippines on these structures
is explored.*

Bi-modal agrarian structures are those in which agricultural
production units (farms) divide into two clearly distinguished
strata in respect of the magnitude of the farms as economic
undertakings, the lines of cropping, their 'mercantility', or
market orientation, and the socio-political situation of the
cultivators who control them. The term 'dualistic' has in the
past been applied to these structures, but it is misleading and
mystifying, since it has been allowed to suggest a society with
two levels of 'civilization': a 'modern' enlightened élite strug-
gling for progress, but held back by an ignorant peasantry. A
closer look at bi-modal structures, however, reveals functional
links of an essentially exploitative kind attaching the two
strata to one another, as will be demonstrated and illustrated
in the course of this book.

Bi-modalism today characterizes the agrarian structures of
most Latin American countries, as national governments,
since independence, have maintained structures implanted
under the colonial regime. These gave to the élites control
over a decisive proportion of land available for agriculture,
both for the purpose of production and as means of generating
a labour force for their enterprises.

A discussion follows of the introduction of the new technology in three countries classified as bi-modal in respect of
their agrarian structures. In the first, Mexico (as has been
already pointed out), the spread of improved wheat technology coincided with the rapid emergence of entrepreneurial
farming out of the aftermath of a period of revolutionary
change involving the decline of the manorial estate, and a
new peasant agriculture built around the *ejidos*, afterwards
abandoned in favour of the encouragement of capitalist
agriculture.

This will be followed by discussion of the introduction of HYVs in Tunisia and the Philippines, the one owing its bi-modal character largely to French colonialism and the other to an older Spanish colonial order.

I. MEXICO

The attempts by Spanish colonizers to develop agricultural production in Central and South America, following the immediate depredations of the Conquest, were for the first century built around the institution of the *encomienda* or entrustment. This institution implied a system of direct rule of local populations, the purpose of which was the stabilization of the newly-conquered colonies, maintaining some of the local elements of indigenous social structure and seeking acculturation and social control by means of Christian indoctrination and the destruction of native religion. At the same time, it was the means whereby the Spanish sovereign rewarded *conquistadores* and other loyal subjects, delegating to them the substance of seigneurial power, with rights to exact compulsory labour and tribute in kind from the native populations entrusted to them, for the benefit of the Spanish Crown and the *encomendero* himself.

In the last decade of the sixteenth century, reforms were carried out whereby *de facto* control and possession of lands formerly in *encomienda* could be transformed into private property by those to whom they had been entrusted. Other lands, which remained Crown property, were set aside as reserves for the subsistence of the local native population. Labour for the new proprietors in excess of that which they could bind directly to their property was provided by the institution of *repartimiento*, a type of *corvée* duty levied on the native population of the reserves and by other contrivances obliging natives to labour in public and private concerns as status-duty derived from the terms of their subjugation to the Spanish Crown.

The unsatisfactory nature of a constantly changing labour force impelled the emergent proprietary class to build up a permanent resident work-force. A class of predial dependants attached to the estate by a variety of institutional devices and bonds was established. The essential condition for creating such a class was to engineer the exclusion of a section of the rural population from access to land and its resources. Officially, subsistence lands of native communities were reduced by the sale of reserves or parts of reserves deemed to be underpopulated. Unofficially, estates encroached on these lands with the connivance of local officials. Those who lost their birthright in this way and could find no alternative livelihood sought the usufruct of subsistence lands on the estates of Spanish and *criollo* landowners and in return pledged themselves to labour-service for the proprietor, or in some cases to the delivery of a part of the product.

It was in this way that during the eighteenth and nineteenth centuries a dominant class of landowners (at first only Spaniards, but later native born persons or *criollos*), gained control of most of the accessible and useful land. And since they exercised political power locally and nationally as well, they were able to use their control of land to create a supply of bonded or wage labour for estate and industrial labour purposes.

Plantation Era Begins

Concurrently with the growth of manorial estates, in areas of special suitability for intensive production of export commodities, i.e. the coast of South America and the Caribbean Islands, 'plantations' developed. The plantation was one of the earliest responses to the growth of a European market for exotic tropical products with a high scarcity value in the temperate climate. The production and processing in preparation for transhipment of sugar, cotton, coffee, and cocoa were typical plantation activities, and it came into existence essentially to produce for the market and make profits for its owners. Internal arrangements were derived from the rational pursuit of this end.

The situation of the plantations was determined by the fact that certain areas not only offered soils and climate favourable to the staple in demand but also transport facilities to the consumer centres located near the coast. In most cases, the areas in question (frequently hot coastal plains) did not offer locally recruitable labour, thus establishing the purchase of slaves as the typical means of securing it, to be followed by indentured labour, and latterly by wage labour as the population grew and the land squeeze continued.

With these antecedents, the plantation was a capitalistic type of agricultural organization in which a large number of unfree labourers were employed under unified direction and control in the production of a single crop. It was characterized by a sharp separation between worker- and employer-class emphasized by cultural/ethnic differentiation. The plantation became a feature of European colonialism not only especially in Brazil and the Caribbean Islands and the Pacific coast of South and Central America, but of certain African and Oriental countries as well. With the intensification of pressures from the industrial countries to incorporate ever new areas and populations within the international market economy, and the development of national dependent capitalist economies and urbanization in Latin America during the last thirty years, manorial farms have become commercial enterprises of varying efficiency, losing their manorial quality.

Side by side with manorial estates and *haciendas*, there have been the reserves of *indios* and *ejidos* (commons) legally established by the Spanish or Portuguese colonial powers and later transformed by purchase into townships and dispersed neighbourhoods of small populations; the fragmentation by inheritance of large estates; land grants to disbanded soldiers or liberated slaves; the spontaneous colonization of population frontiers, river banks, and new highways by squatters; government internal colonization and land settlement schemes for landless people and for recent European immigrants.

If we consider Latin America for a moment, a study of land tenure done in the 1960s published data about the land tenure status of families in agriculture in five countries, which yields the results as shown in Table 4.

In the 'estate' sector of these countries, at least two thirds of the agricultural land is controlled by a small number of estate operators, on whom the landless, numbering between one quarter and three fifths of the agricultural population, depend as service, share or cash tenants, as permanent wage- or salary-earners and as day-labourers. In the smallholder

TABLE 4

Tenure Status and Value of Production of Agricultural
Families in Five Countries
(in percentages)

| Country | Agricultural Families | | | Land Occupied | | Value of Production | |
	Estate Operators	Small-holders	Landless	Estate Operators	Small-holders	Estate Operators	Small-holders
Brazil	14.6	23.5	61.9	93.5	6.5	78.7	21.3
Chile	9.5	40.8	49.7	92.6	7.4	80.0	20.0
Colombia	5.0	70.3	24.7	72.8	27.2	47.8	52.2
Ecuador	2.4	63.1	34.5	64.4	35.6	40.7	59.3
Guatemala	1.6	73.4	'27.0	72.3	27.7	56.4	43.6

Source: Pearse, 1966.

sector, between 20 and 60 per cent of total production was accomplished by between 23 and 73 per cent of agricultural families.

In 1940, almost half of the cultivable land of Mexico was in the possession of *ejidos*, as will be seen in Table 5, and a further 10 per cent of it belonged to smallholders owning 5 hectares or less.

TABLE 5

Distribution of Cultivable Land and Production,
Mexico, 1940

| | Cultivable Land | | Value of Production | | |
	ha (000)	%	Pesos (000,000)	%	$ per ha.
Private proprietors	3,045	42	295	39	96.9
Ejidos	3,548	48	392	51	111.4
Smallholding proprietors	749	10	75	10	100.1
Total	7,312	100	392	100	104.2

Source: Eckstein, 1966.

The Cárdenas government had been able to create the majority of these *ejidos* either by restoring land to communities from whom it had been taken in the pre-revolutionary period (pre-1910) or by the expropriation of large commercial farms (many of them owned by foreigners), and their establishment as the joint property of their former labourers. As a result of this policy, the number of landless labourers in the rural work-force had dropped from 68 to 36 per cent. Normally an *ejido* was an area of farmland assigned to a village community or other collective of assignees. However, in cases where large expropriated farms had been commercial enterprises enjoying the economies of scale, these had been transformed into producers' co-operative farm enterprises.

Disintegration

From 1940 onwards, this whole apparatus was in the process of dismantlement. Public investment now flowed into infrastructure, transport, and credit for the private sector, as we have related in Chapter I, and *ejido* cultivators found themselves in a weak and deteriorating position. The post-Cárdenas governments left the agencies connected with the Ejido Bank to drift, with limited budgets and dispirited personnel, towards incompetence, and in the worst of cases, deliberately utilizing them to sabotage the efforts of groups of farmers associated with the political left. Credit was delayed, inputs of poor quality delivered to *ejido* clients, and technical assistance virtually abandoned. Attempts at the regional organization of co-operatives to lessen dependence were crushed by force and by the mid-1950s the largest and best-endowed *ejidos* of Mexico had been broken into hundreds of bickering fractions, each trying to salvage what its members could from the wreckage (*Hewitt de Alcántara, 1976*).

By the time the authorities showed any interest in introducing the new technology to cultivators in the *ejido* sector, the process of disintegration was far advanced. The occasion of the authorities' interest in the yields obtained by *ejidatarios* was the 1953 drought and the return of rust-disease, so that a nation-wide campaign had to be initiated to multiply high-yielding disease-resistant seeds. By this time, however, there was no coherent *ejido* sector organization through which the cultivators could be mobilized. Without any warning, *ejidatarios* who received credit from the Bank had *Lerma Rojo* seeds dumped in their fields, with no fertilizer, no insecticide, and no explanation.

For the next ten years, the *ejido* cultivators were involved in the Green Revolution in the worst possible conditions. While the scientists struggled to find wheat varieties that were resistant to several new kinds of stem-rust, as these appeared, the operations of *ejido* cultivators were tied to the Ejido Bank, a rigid organization politically disposed against their development, for credit, supplies, and information.

For the first three years, it appears that no fertilizer was delivered, and the seeds they received are reputed frequently to have been old stock, vulnerable to the reappearing rust varieties on which the bank did not wish to lose money. As a consequence of the shortcomings, indifference, and venality of officials of the organization, which monopolized their nexus with the source of the new technology and its inputs, combined with the disorders of decay of the *ejidal* system, the wheat yields of the *ejido* sector dragged behind those of the private sector between 1953 and 1965 (*Hewitt de Alcántara, 1976*, p. 211), in contrast with the 1941–5 period, when the yield figures for the *ejido* sector matched and sometimes surpassed them. And behind those figures for *ejidatorios'* wheat yields lies the revealing and highly significant process of the distortion of institutional structures by market forces, but with an outstanding contrived advantage in favour of the entrepreneurial class.

Mainly as a result of the failure of the Ejido Bank to function as a credit and technical assistant provider, ever increasing numbers of *ejido* members were obliged to rent out their lands clandestinely to entrepreneurial cultivators, accepting occasional wage work on their own lands and a rental that could be artificially held down because of the threat to the *ejidatario* posed by the illegality of his act. By the mid-1960s, as many as 80 per cent of the *ejidatarios* of the Yaqui Valley had abandoned control of their land.

An exception to this process in the form of a few surviving collective *ejidos* provides some further illustration of the forces at work within the bi-modal agrarian structure.

Hewitt de Alcántara (1976, pp. 214–34) provides us with a detailed account of one such producers' co-operative, to which the reader is referred. The Quechehueca Collective Society had excellent leadership and by its organized strength it was able to invest in improved productive equipment and machines, although it is noted that the insistence on consensus in taking decisions about cultivation practices considerably slowed down innovation.

The Society was able to defend itself against the bureaucracy, and it found ways of handling the typical problems of producers' co-operatives connected with the distribution of income. It also established some social services and facilities for consumer credit. In the course of its development it suffered a considerable loss of membership by withdrawals. But the advantages of membership are to be seen in the present modestly affluent situation of its members. It is ironical but perhaps inevitable that the collective, originally formed by poor labourers, but with backing from the authorities, now makes use of wage-labour of non-members for most of the tasks of cultivation.

II. TUNISIA

The Green Revolution episode in the fortunes of Tunisian post-colonial development is a matter of lively interest since the new technology was introduced and propagated in an effort to improve productive methods while at the same time building an agrarian structure rid of a crass bi-polarity inherited from French colonial rule.

The Tunisian case has many parallels with the Mexican one in the sense that the introduction of the new technology coincided with withdrawal from a populist policy that had attempted to strengthen peasant agriculture through the establishment of co-operative units of production and agricultural services for the common cultivator.

Tunisia acquired a bi-polar agricultural structure as a result of the intrusion of large-scale commercial agriculture, mainly French, in the northern plains during the colonial period, which ended in 1956. This intrusive agriculture was devoted chiefly to the monoculture of grain, and was highly mechanized, having achieved a level of one tractor per 45 hectares and a combine harvester for every 160 hectares by 1930.

In the 1930s also, *Florence Aurore*, an improved variety of soft wheat, had been introduced and its use widely diffused on the colonial estates, reaching a yield of 12.5 quintals per hectare as compared with 3 quintals of peasant hard wheat. However, in spite of these modernization trends, the agricultural activity of the *colons* had had two harmful consequences. Its expansion put excessive pressure on peasant agriculture by taking over much of the best lands and confining traditional agriculture to marginal lands (as happened during and after the Spanish colonial period in Latin America). On the other hand, its extractive nature was damaging to the quality of the very light soil (*Hauri, 1974*) on account of excessively deep ploughing, the destruction of humus, and the failure to maintain or adopt conservation practices by means of suitable crop rotations.

At the same time, the combination of pressures on peasant agriculture, plus the use not only of tractors but of combine harvesters, had created unemployment. Thus the colonial occupation of Tunisia ended, leaving severe problems of deteriorating soils, a land tenure situation characterized by excessive inequality as well as the complications of customary feudal-like relations, and of widespread unemployment. Following independence, plans were elaborated for carrying out a combined programme of structural change and technological development.

According to the Tunisian politican Ben Salah, the master task of development of the new nation was that of knitting together again a country split into two by its colonization, a cleavage expressed in the polarization between French agriculture and that of the marginalized peasantry. Development planning involved investment in irrigation, in fertilizer, in agricultural credit, and the technological improvement of production. But the question was posed: by what channel should these new elements be brought to the point of creating wealth? The task of modernizing the fragmented and marginalized peasant holdings was a daunting one, while the channeling of this investment to the large properties would lead to the further exploitation of the peasantry. 'The peasants can't afford to buy the water so they sell their next three years' harvests in advance to the neighbouring large proprietor in order to get a few drops on account, as secondary utilizers, or else they must go without.' (*Nerfin, 1974,* author's trans.)

The new institution designed to undertake the knitting-together function was the agricultural producers' co-operative, whose membership was composed of former workers and technicians on the expatriate-owned estates to whom were added some of the peasant cultivators who worked on their fringe, some of whom entered with their land.

The programme establishing the new productive structure began in 1962 and by 1968 had made limited progress only. In 1969, the leaders of the programme attempted to accelerate and extend its operations, but there were powerful political reactions and the main protagonist of the policy of structural change was driven from office. The programme was dismantled, and individual members of the agricultural producers' co-operatives (UCP) were obliged to withdraw and return to the individual cultivation of their lands. Indeed, the political volte-face that overtook agrarian policy in 1969 was not so different from that of 1940 in Mexico, discussed above. As a result of it, the idea of a programme combining structural transformation with technological innovation was maimed, and an attempt was made to introduce the new technology in the midst of a disconcerting reversion to bi-polarity. The conflicts and contradictions of which this reversal was symptomatic undoubtedly contributed to the failure of the programme, notwithstanding the agronomic promise of the Mexican HYVs.

A determined attempt was made to introduce improved technology in grain production. As a part of this policy, in 1966 the Tunisian National Agronomic Research Institute began experimenting with Mexican wheats. These varieties have given their best results in Mexico under conditions of controlled irrigation, but for dry agriculture, too, they offer certain advantages. Their shorter vegetative cycle makes it possible to complete the harvest before the great droughts are under way. Thanks to their non-photosensitivity, they can be grown in a variety of latitudes, and they had acquired rust resistance. They are, however, soft wheats rather than hard, and this in Tunisia signifies that they are not acceptable as a rural subsistence crop. Hard wheat is eaten in the form of *couscous* by country people and the majority of town people as well. Soft wheat is cultivated to be eaten in the form of bread, an urban taste of increasing importance.

One of the questions posed here has to do with the role played by the package and its 'fit' in the relative failure of the experiment. Characteristics of the Mexican wheats are given below. Introduced experimentally by the Institut de la Recherche Agricole de Tunisie, the Mexican HYVs were submitted to field tests in 32 different lots in UCPs of North Tunisia, and the results were very favourable in comparison with local varieties, as is shown in Table 6.

TABLE 6

Experimental Wheat Yields in Field Tests, Tunisia, 1967–8
(quintals per hectare)

		Dry	Irrigated
Inia	(Mex)	26.4	36.5
Sonora	(Mex)	23.0	33.1
Javal	(Mex)	21.0	29.4
Tobari	(Mex)	22.0	36.4
Florence Aurore (soft)	(Tun)	17.0	20.8
Mahmoudi (hard)	(Tun)	12.0	17.9

Source: Hauri, 1974.

At that time the average yield for hard wheat was 5 quintals and for soft 7.5 quintals per hectare. The Mexican wheats, therefore, with the necessary application of nitrogen and phosphorus and cultivated according to certain modified practices, offered the possibility of a very significant jump in yields.

Leaving aside the cost of labour, the main cash cost of production for local soft wheat per hectare was 4,300 dinars for 1 quintal of seed, from which a crop of 7.5 quintals could be expected. With a change to the Mexican wheat, the cost of seed rose to 6,500 dinars and to this must be added 9,500 dinars for the recommended amount of nitrogen and phosphorus. This addition to costs of 11,700 dinars that the Mexican grain required would be covered if yields were raised 3 quintals per hectare, from 7.5 to 10.5 quintals. On this reckoning, a yield of 17 quintals per hectare from the Mexican variety would have doubled the profit as compared with the local variety.

Irène Hauri gave an account of the introduction of the HYVs in the four successive seasons of the experiment (*Hauri, 1974*) in the Development Region of Pont-du-Fahs. During the crop-year 1968/9 Mexican seeds were issued to selected individual cultivators and to UCPs to plant some 6,000 hectares. Results showed a high degree of variation, with yields averaging 15 quintals per hectare.

In the second year, the aim of the programme was the sowing of 150,000 quintals of Mexican seeds by UCPs, large cultivators, and peasant cultivators in the 10–100 ha. class. At first, there was very little sale of seeds so the promoters decided to sell them for one quarter of the price down and the remaining three-quarters at harvest time. However, by the time this offer was made, many people had already sown. Nevertheless, to the smaller cultivators this offer came as a windfall, since many were facing economic difficulties after withdrawing from the UCPs, and so they grabbed the opportunity of getting seeds at only a quarter of the price.

There were also difficulties in obtaining fertilizer since hitherto it had been distributed from the capital and there was virtually no infrastructure established for local small-scale distribution, while credit for herbicides and fertilizer arrived late. However, the weather was good and average yields reached 20 quintals with some larger cultivators achieving 40 quintals.

Thus the use of HYVs was a technological success and the difficulties were surmounted by UCPs and larger cultivators. But arrangements for individual peasant cultivators were quite inadequate, and only 50,000 of the intended 150,000 quintals of seed had been distributed.

In the following crop-year 1970/1, impelled by the excellent yields obtained, promoters of the new technology pushed the target up to 100,000 quintals, with special arrangements made for credit in kind to the peasants, while bank credits for large cultivators were raised to cover 30–40 per cent of their expenses.

The harvest was a catastrophe for the programme and peasant cultivators reaped on the average only 3 quintals per hectare as against 6 to 7 quintals for the traditional hard variety. Sowing was late as a result of late rains, and peasant husbandry fell short of the precise requirements of the Mexican varieties: the peasants mostly planted too deep for fear of drought, and they failed to keep the weeds down. But most important of all, the seed distributed as credit in kind consisted of unsold and untreated leftovers from the previous year, and largely failed to germinate. Disillusionment was complete. In the fourth crop-year 1971/2, a few larger cultivators managed to plant HYVs using their own seed and buying fertilizer. Some of the remaining UCPs also used the new technology, and the total area sown reached 4,800 hectares, but the peasantry dropped out, and thereafter returned to the accustomed hard variety.

Can it be said that the Mexican wheat was 'built' for areas of reliable water control and was an inappropriate package for Tunisia? Field experience showed that the Mexican wheats performed well even in dry conditions, but only when cultivated and treated with fertilizers in a certain way. This means that they were appropriate for a limited sector of the Tunisian cultivators.

The Mexican wheat was especially suitable in Tunisia because of the early maturing attribute which made it possible to gather in the harvest before the worst of the dry weather set in. But this required an additional condition: that the delivery system should get seeds to the cultivators on time, and this the package delivery system failed to do.

We mentioned earlier (p. 12) the oversimplification involved in the concept of a package. The same may also be said of the *delivery system*, which suggests a neat apparatus, but refers to the *whole problématique* of the control and distribution of resources in society. A delivery system, if it is to handle essential factors of production for agriculture, will reflect the

market pressures and forces of the society in which it functions. In the attempt made by the HYV programme sponsors
to reach out to the poorer peasants in the seasons 1969–70
and 1970–1, registration of applicants was put into the hands
of the Union National des Agriculteurs, whose primary motivation was the search for political support, and which did not
therefore apply realistic criteria in recruitment. The list
drawn up by this body was then submitted to the traditional
sheikh or local boss for confirmation of the candidates'
property rights. However, since property rights were fluid,
and since the sheikh was responsible for the issue of property
certificates, the task was carried out in a highly personalistic
manner and the resulting list bore the imprint of the sheikh's
likes and dislikes.

The management of the list by these two political forces
is obviously not unconnected with the low credit repayment
rate. But in addition to this political handling, the applications had to pass through six different offices. Delay was
even greater because many cultivators applied only when
they found that no hard wheat seeds were available—HYV
seeds were therefore planted late, in the worst fields in the
absence of anything better. But they were attractive because
payment was deferred and, it was hoped, avoidable.

The Mexican wheat was marketable. However, it lacked
the quality of 'eatability' for the small cultivator who was
traditionally accustomed to *couscous*, which could not be
made from Mexican wheat. While the cultivator who is predominantly commercial in orientation is motivated to achieve
tangible monetized profits, the small cultivator in Tunisia,
for example, puts the achievement of a secure livelihood first,
measured by a minimum of fat sacks for his family's food.
The Tunisian peasants were prepared to risk indebtedness to
the government, but not to take chances with their subsistence
basket—their few hectares of local hard wheat.

The HYV package was in the circumstances inadequate as
an approach to peasant agriculture, which was still self-
provisioning. The peasant had no savings to invest and no
guarantees for obtaining credit—the conditions for HYV success seemed automatically to exclude the mass of small cultivators who had nothing in hand with which to work. A

package for them needed minimal production costs, and maximal survival qualities to meet climatic excesses.

If the scientists' work of fitting the physical package to the ecology is to be successful, it must be oriented by the more ample assignment of fitting a new productive system to a society in flux.

III. PHILIPPINES

The complex of programmes executed by the Philippines government to raise rice productivity and so dispense with large-scale rice importation was subject to a variety of conditioning influences, of which the importance of three must be underlined: the system of share-tenancy to which the majority of rice producers are subject; a type of bi-modalism that separates food crop production (rice and maize) from the large-scale production of commercial and export crops; and the highly political nature of the rice supply and its price.

The prevalence of share-tenure cannot be fully understood apart from the colonial background. In the late 1880s the Spanish rulers passed laws calling on landholders who occupied land by virtue of customary rights to obtain individual titles to ownership, but the opportunity was taken mainly by the *caciques*, whom the Spaniards had maintained as local agents of their rule and fiscal control—and also by clergy from the church and the religious orders. The result was that a large section of the peasantry found that the land they were accustomed to cultivate now belonged to *caciques* or to the clergy, leaving them under the threat of eviction or in the status of share tenants. As a result of this land-grabbing, and the rupture of the traditional order, the final years of the Spanish colonial regime were marked with agrarian conflict.

The Americans occupied the country in 1898 and faced the problem of issuing land titles to 400,000 peasants. A system of land registration was introduced but the problem of titling was not settled, and by 1902 only 4,000 titles had been granted.

During the present century, other factors have weakened the position of the working peasantry. Large-scale acquisitions of agricultural land, especially for sugar cane cultivation, by nationals and foreigners, have restricted the area available for peasant expansion. By 1938 it was estimated that 346,000 hectares of land had been bought by American, Spanish and Philippine business houses, and rented out on a cash basis for commercial agriculture.

There has also been the expansion of a rural middle class whose members have been able to acquire paddy lands as property for the purpose of investment through their control of commerce and capital. This rising class, consisting of commercial people, government officers, and professionals, appears in particular to have taken advantage of the many sales of large landed properties that had taken place since the 1930s due to threat of peasant uprising and land reform. These properties were divided up and sold in blocks and sizes to suit the savings of these new middle-strata people. The evolution of this petty landlordism is seen in Table 7.

TABLE 7

Landowners and Cultivators in Plaridel According to Size of Property/Size of Operational Holding

Size categories	No. of landowners	No. of cultivators
Under one hectare	64	232
1.0 to 1.99	148	620
2.0 to 2.99	94	329
3.0 to 3.99	44	113
4.0 to 4.99	33	38
5.0 to 9.99	50	15
10.0 to 19.99	26	4
20.0 to 29.99	8	—
30.0 to 39.99	10	—
40.0 to 49.99	1	—
50.0 to 59.99	2	—
60 and over	1	—
Total	481	1,351

Source: Abdul Hameed, field notes, 1973

It is extremely difficult to ascertain the actual state of land ownership in the Philippines today. At the time of the 1960 Agricultural Census, 40 per cent of all farms were cultivated by tenants and a further 14 per cent by cultivators who were tenants of a part of the land they worked. Unfortunately, these figures do not show what proportion of these were commercial tenants rather than traditional and dependent share tenants. However, in the most important and productive area for rice cultivation, Central Luzon, we know that farms operated by their owners or managers amounted to only 23.1, 28.7, 30.5, 10.7, and 24.1 per cent respectively in the five most important provinces of Bulacan, Nueva Ecija, Pagasin, Pampanga and Tarlac. The rest of the cultivators in each province held the whole or a part of their farms in some form of tenancy.

Other data, from the municipality of Plaridel in Central Luzon, and collected from a Land Reform Team by a Global Two researcher in 1973, gave a curious and revealing picture of rice landlordism.

Of the 482 landowners in Table 7 only 20 per cent were cultivators so the overwhelming majority were petty landlords who, it must be inferred, had acquired their properties as investments. The Plaridel population therefore contained these two important groups: 1,350 landless peasants cultivating rice on the lands of some 380 petty landlords, to whom they were obliged to deliver half the crops as rental payment.

The operation of this system will be examined in greater detail.

Persistent Dependence

The political importance of rice must be seen against the search for an adequate and politically feasible development strategy, following World War II. At that time the typical colonial export orientation began to be complemented by a policy of import substitution, making use of input and exchange controls and tax exemptions for certain industries. However, by the 1960s, it was apparent that this policy of development was not bringing with it the loudly proclaimed progress towards widely shared prosperity, and a number of reasons for this were apparent, particularly the failure of the expanding complex of industrial enterprises either to provide sufficient employment for the growing population or to stimulate agriculture by generating linkages that would boost production of food and fibre. Domestic food production, the central element of which was rice, was maintained by a numerous body of cultivators (the majority of whom were share tenants) and a declining proportion of petty proprietors, all of them producers whose poor harvests fluctuated with the weather.

This weakness at the heart of the economy spelt persistent poverty and dependence among the peasantry, and an equally persistent murmur of rural insurrection, while the ups and downs of the paddy (unmilled rice) supply destabilized the price of rice and consequently the whole urban political ambience.

In attempting to explain the motive force behind the far-reaching campaigns for the spreading of the new technology in the Philippines from 1966 onwards, Mahar Mangahas (1974) insists on the close relation between rice production and politics. The post-war political scene was one of factional struggle between rival coalitions of landed and business élite groups, fought out in local and national elections to the music of the new media. Mangahas shows that during the years from 1960 to 1967, decisions to import rice coincided with the occurrence of elections both local and presidential, thus suggesting attempts by the administration to win electoral support. After presenting import figures, he adds:

Any Manila resident will recall the typical propaganda during an election year. Although elections are held in November, activity is stirred as early as January or February. This is very soon after the main rice harvest in Central Luzon and in most other regions, and so rice prices at both the farm level and retail level are seasonally low. Nevertheless, as early as this there will be reports and soundings in the press about an impending rice crisis. Some of this propaganda may stem from the administration party, since it wished imported rice to arrive during the peak of the campaign, from July to November. It realized that rice prices were seasonally high during the campaign period, and the idea was to try to dampen the seasonal movement and prevent the opposition from capitalizing too much on high rice prices during the campaign period. The opposition party also had cause to contribute to the propaganda thus criticizing the administration party for an ineffective agricultural development policy, and in addition preparing the minds of voters for a rice crisis later on in the year.

In spite of the use of this cumbersome electoral paraphernalia, rice imports do not seem to have been very effective in bringing down prices, nor did any president win re-election to office until 1969. But, however sceptically one may regard Mangahas' correlations, there is little denying that rice, its price, its importation, and the control of rice supplies were highly political issues. It was not a simple question of the price of rice: equally important values were involved appealing to populist sentiments, and a national aspiration to achieve freedom from political leverage.

From the point of view of good housekeeping there were persuasive reasons for the administration to launch technology-raising campaigns for rice producers, in view of the low level of existing technology, the loss of impulse in the expansion of the area under irrigation, and the presence and plant-breeding success of IRRI.

Rice had been an important food crop under the Spanish colonial regime. Surpluses were required for Spanish settlements and as food supply in those regions and islands of the colony specializing in the production of some crop that did not iself provide local subsistence. Rice was important for the Japanese, too, during their occupation of the islands and they commandeered it as a direct tribute, against peasant resistance. During the 1950s the government set up a board for the distribution of certified seed, and the government's Rice and Corn Authority began acting as a distributor of fertilizer towards the end of the decade.

During this period, investment in irrigation contributed to improved yields and in the 1960s the Authority assumed responsibility for price policies, as rice production became increasingly profitable. However, there was only a minor increase in yields.

Careful Planning

The first general release of HYV seeds was in 1966. In the same year when new varieties were released to seed-farms for multiplication, yields ranging between 1.6 and 6.4 tons per hectare were obtained, as compared with the national average at that time of about 1.3 tons per hectare. In some ways the strategy used for the introduction of the IR-8 seeds and the appropriate technology were a model of rural development planning. In that year, the Rice and Corn Production Consultation Council was revived with the President of the Republic in the chair. Participating agencies included those responsible for irrigation, research, extension, price support, credit, co-operatives, land reform, and soil analysis.

The bringing of the new seed to the cultivators and their persuasion to make use of it was carefully planned. Areas favourable to these experiments because they had irrigation, warehousing, rural banks, and agricultural supplies were chosen. Essential supplies were provided as a 'package' or 'kit', along with printed instructions. One thousand six hundred production technicians were given a few weeks' training and their numbers were later supplemented by more than 700 specially trained farmers. Each technician was supposed to cover 300 hectares of HYVs and he would meet with the village council to select 'progressive' farmers willing to experiment with the new technology.

This extension effort was accompanied by further irrigation works that raised the irrigated percentage of rice lands from 31 per cent in 1963–5 to 42 per cent in 1968–70 (*Barker, 1971*). Some idea of the performance of rice cultivators can be obtained from yield figures provided by the FAO Production Yearbook. Yields for the years 1961–5 averaged 1,257 kg. per hectare and continued as follows: 1967: 1,440, 1968: 1,330, 1969: 1,680, 1970: 1,720, 1971: 1,571, 1972: 1,483, 1973: 1,628, 1974: 1,599, 1975: 1,737, 1976: 1,808.

What kind of explanation can be found for these figures showing an upward trend in yields? One very obvious one is the weather, and all that low rainfall or excessive rainfall and storms may mean in terms of plant growth, disease, harvesting problems, etc. Undoubtedly, most of the heights and depths can be seen to be related to weather conditions, i.e. the droughts of the mid-1960s, and typhoon of 1972, and the superb weather enjoyed for the 1975–6 crops.

A second factor offering further explanation is irrigation. During the ten years contemplated, there was a steady expansion of the paddy areas irrigated. The possibility of controlled water applications at appropriate stages in the growth of

HYVs is crucial to the achievement of the high yield effect, and in non-irrigated areas it is only in exceptional circumstances that moisture is available to the plant at the 'ideal' time.

In fact, the bold quantification of the national average paddy yield as x kg. per hectare is usually arrived at simply by dividing the amount of the gross product by the number of hectares sown to a particular food-grain, an umbrella that hides more than it reveals save the reasonably secure indications of yield levels and trends over periods of seven years or more.

However, it does not serve the purpose of the present undertaking to concentrate too exclusively on average yields. It is more important to ask whether the present agrarian conjuncture and the measures in hand to manage it offer both prospects for adequate growth of total rice production and a rewarding livelihood for rice producers. In order to provide an answer some further elucidation of what tenancy in rice production signifies is required.

Tenancy System

The relationship between tenant and landlord, which lies at the centre of contemporary rice production in the Philippines, has evolved out of a customary patron–client dependence institution known as *kasamahan*. The most important features of the context in which this evolution has taken place are the rapidly rising rural population and potential work force, increased demand for rice, commercialization of the social relations of production, and the introduction of industrially produced inputs, tools, and power into agricultural production.

What is referred to as a 'patron–client relationship' is, of course, a widely known relation of dependence of the greatest importance in feudal and post-feudal Europe, and in one form or another still central to agrarian relations (and special ones) in Asia and Latin America. In its most common agrarian form it implies control of land by a particular social class whose members are able to exact services, rents and tributes in return for the concession of usufructuary rights. Two recent authors who examined patron–client relations in South-East Asia coupled its existence with the absence of effective impersonal guarantees such as public law for physical security, property, and position, and the inability of either kinship units or the traditional village to serve as effective vehicles of personal security and advancement. Such a condition has been common until recently, especially in former Spanish colonial territories, and their manorial arrangements express a structural accommodation to this situation.

So long as landlords have difficulty in securing an assured labour supply, there is likely to be benefit from both the landlord and the tenant in the compact between them stipulating duties owed materially to one another. According to *kasamahan* as practised in the cultivation of local rice varieties, the landowner offered the tenant use of the land and provided the seeds and advances necessary to cover cultivation expenses. The tenant or *kasama* produced the crop with his own animals and tools. Division of the harvest between landowner and tenant took place after subtraction of the expenses borne by the former; that is to say, cultivation expenses were shared equally.

Landowners were also supposed to make advances of rice to the *kasamas*, if and when they entered his service, and an annual advance without interest to sustain his family up to havest-time, and to lend him money when he was in special need of it. In return the landowner could call on the *kasama* to run errands for him and provide certain personal services such as helping with house repairs and gathering fuel for the estate household. It was also customary for the *kasama* to bring fruit from his garden for the landowner.

Formerly, the relationship between the landowner and tenant was not usually a temporary and insecure arrangement, but had the character of a permanent tie, in which the tenant enjoyed a certain degree of basic security as regards his livelihood and protection in his dealings with the wider world, at the cost of considerable economic exploitation in which his dependence is maintained through indebtedness. That both substance and recollection of the customary relation linger on in places is demonstrated by current field research.

Quoting a study by one of her students, Gelia Castillo (1975, pp. 296 f.) lists five expectations about landlords' duties that were widely shared by tenants. The landlord was expected to (a) provide the tenant extra emergency funds aside from operational farm expenses, which are non-deductible from the harvest; (b) praise his tenants for a good harvest; (c) give free rice and other food supplies to augment families' supplies; (d) let tenants decide how to cultivate the land and what variety to plant; (e) be strict with tenants who are not his friends or relatives. Personal field inspection by the landlord is not regarded as necessary.

The increasing importance of interest on loans by the landlord to the tenant as operating capital became more important is noted before the Green Revolution period. Takahashi (1971) describes the system as it operated in his village in 1963–4 as follows:

Landlords are the most common sources of funds for tenant farmers, and described themselves in such an expression as 'we are like fathers of tenant farmers' . . . it is common for tenant farmers to repay with principal and interest in palay, and even when they repay the principal in cash, they usually repay the interest in palay. The ordinary rate of interest for the crop season amounts to 2–3 cavans of palay or 30 to 45 pesos on a principal of 100 pesos. Thus the annual rate is 60 to 90 per cent. When the principal is repaid in palay the price per cavan usually is estimated about one peso lower than the producers' price. When rice is borrowed, the common practice is to repay three cavans of palay on a principal of one cavan of cleaned rice (equivalent to two cavans of palay), or at the rate of 50 per cent interest per half a year. (p. 88)

(See also *Bernal, 1971*, p. 92, and *Mears and Agabin, 1971*, p.36.)

Unequally Yoked

On the basis of these discussions, there seem to be two factors in the tenants' situation that have contributed to the growth of yields of paddy. As with very small cultivators in many other parts of the world without alternative employment opportunities, the high marginal utility of their incomes has impelled them to put more labour into their cultivation practices and to use the land more intensively. The marginal inferiority of yield on the part of the owner cultivators is attributed to the fact that their improved economic and social situation restricts the use of family labour and especially that of their children, who aspire to jobs and status requiring schooling more than that afforded by the paddy field. In the same way, perhaps, the owner operator is likely to have more diverse opportunities for investment.

The tenant, on the other hand, finds himself unequally yoked in co-entrepreneurship to the landlord, who may serve not only as an immediate source of credit but with his superior education and urban connections is well placed for receiving and passing on useful technical information, and getting the most out of government services and facilities. Indeed, it was found common for landlords to take decisions about which seeds to use and how much of what fertilizer to apply, on the basis of printed information to which they have easier access than their tenants. They also finance one half of the bought inputs, and serve as bankers for the other half, which will be repaid at harvest time.

In many cases, it was also found that landowners had become service entrepreneurs as well, and had invested in farm machinery that could be hired by the tenants to replace human labour, if they judged such a replacement expedient. Most tenants, therefore, through the entrepreneurial participation of the proprietor, seem to have avoided the contractual inferiority of the poor man in the factor market, his lack of political 'clout' in getting his share of government services, and the illiterate man's difficulty in ensuring the reliability of technical information. In all these respects he benefits from the backstopping if not the direct help of his 'co-entrepreneur'.

But the price he pays is a high one, namely excessive interest on his loans, and debt dependence. Indeed, it is the struggle between the two entrepreneurs at the centre of each farm within the modern version of *kasamahan* that expresses the form taken by agrarian class conflict between a sector of those who control both land and capital, whether on a large or small scale, in the rice lands of the Philippines, and the *kasama*—the working cultivator.

Kasamahan is no longer a society for mutual benefit between albeit unequal partners. As two other observers have noted, the position of the landlord has lost the legitimacy the tenant once felt it to have and which caused him to experience the sentiment *utang na loob*, or 'sense of gratitude'.

The present transitional character of the institution of

kasamahan is precisely illustrated by the results of a study of landlord behaviour (*Bernal, 1971*), which found that 84 per cent of them fell into what the writer called the 'modern non-paternalistic type'. They were

. . . people who have grown away from the farm and have been in contact with the industrializing sector. They tend to look at farming as another business that ought to be profitable and (they) would deal with their tenants in the same manner as their business associates. . . . The prevalence of this type reflects an opportunistic combination of what is desirable from the old and the new ways of dealing with the management of rice farms. To utilize paternalism in the interest of increasing productivity represents a very calculated approach on the part of the landlords.

Two very important issues remain to be noted: in order that investment in irrigation bears fruit in improved yields, planned works of land improvement are required to get the moisture to the crops. Local investment in labour and hire of machinery requires the collaboration of landlord and tenant, which their present ambiguous relationship makes difficult. The other point is, of course, the continuing threat of instability posed by the dissatisfaction of the tenant class. A number of important areas of the Philippines have histories of several decades of agrarian unrest, and the legitimacy of the grievances of the *kasama* in relation to control of the land have been recognized by the enactment of land reforms transferring control to the cultivator and providing official sources of credit.

Referring to the land reform measures announced in 1972, soon after the declaration of the actual law by President Marcos, a feature article of 22 September 1976 in *The Times* (London) asserts that at the time of writing four years later, only 16,410 tenants had achieved effective control over their land under the terms of the law by working out a price for the land in negotiation with their landlords and had begun to pay the price in installments. But much though the government would like to count on the support of a contented peasantry, loss of support from the new middle strata who consider petty landlordship of rice lands a secure investment would be a high price for the government to pay.

It is unlikely that land reform will achieve any important

transformations. Perhaps more important for the future is the new attempt to encourage industrial enterprises employing urban labour to make arrangements for the production of sufficient rice to meet the domestic needs of the work force.

V PROMOTION OF THE NEW TECHNOLOGY

Reports are presented on the introduction of HYVs in India, Indonesia, Sri Lanka, and Malaysia to overcome 'stagnation' and 'low productivity' in peasant agriculture. Special attention is given to the way the different programmes have been organized and administered.

The Green Revolution Strategy

The four programmes for the introduction of the high-yielding varieties of wheat and rice in India, Indonesia, Sri Lanka, and Malaysia, presented in this chapter, are all serious attempts to transform food production in societies where agriculture was clearly dominated by peasant production, that is to say, by small farms in which the main labour input came from the cultivator himself and his kin, where the physical elements of production were the products of the farm and its neighbourhood and in which most of the product was consumed by the cultivators and the complementary landless labourers who participated in production in return for wages in cash and grain.

These programmes may be thought of as a variety of different attempts at realizing the 'Green Revolution strategy' for overcoming what was identified as 'stagnation' and 'low productivity' in the peasant sector. In respect of the origins of this strategy, a number of writers have put their finger on Theodore Schultz's (1968) book, as the source of the theory. Shigeru Ishikawa (1970) neatly sums up Schultz's hypothesis as follows:

1. The initial state of agriculture in the developing countries is assumed to be a state of stationary equilibrium which is arrived at over a long period on the basis of the constant state of the arts and the constant state of preference and motives and which is designated as 'traditional agriculture'. In traditional agriculture, people are efficient but poor because (1) there are few significant inefficiencies in the allocation of the traditional factors of production, but (2) since the rate of return to the additional investment in the traditional factors of production is very low, there is no inducement for net investment and, hence, no net saving.

2. The dynamic force for transforming traditional agriculture into modern agriculture comes from new factors of production, which consist of the following two parts. One is 'modern (material) inputs' which embody the knowledge of established scientific theories and principles pertaining to plants, animals, soils, mechanics and so forth (especially the genetic principle underlying hybridization). They are highly productive and cheap (relative to return). The other is skill and other capabilities of people required to use such material inputs successfully.
3. The strategies developed in this model are essentially equivalent to an exploration of the optimal allocation of social capital among the various means for making farmers accept modern inputs. Items of the resulting investment are, among others: research and development of modern inputs, the distribution of the findings thereof to farmers and schooling of farm people.

It follows that in order to create favourable conditions for the kindling of a dynamic development of peasant agriculture, favourable conditions would have to be created from outside the 'traditional' agricultural sector that would lower risk and reduce or compensate for uncertainty; then the main factor in adoption would be the rate of profit.

Schultz saw the problem of the niggardliness of productivity of traditional agriculture as essentially one of investment, and rightly pointed out that the economist of that period who studied growth had concentrated on industrial growth, despite the fact that in low-income countries the agricultural sector was predominant. His approach treats agriculture as a source of economic growth, and the analytical task is to determine how cheaply and how much growth can be realized by transforming traditional agriculture into a more productive sector by means of investment. Schultz's readings also led him to the conclusion that traditional cultivators were responsive to prices and inclined to be profit-seekers.

To the foreign aid pundits these views gave powerful academic support for development policies directed towards agriculture and built around the popularization and diffusion of the new high-yielding varieties of wheat and rice seeds, which could be multiplied biologically at geometrical rates of progression, thus conserving their low cost. And, in particular, the explanation and prescriptions offered by Schultz were justified within purely economic reasoning, and did not involve 'institutional factors'. For this reason, technical aid missions could approach governments with development policies that neatly side-stepped the awkward and subversive ghosts of land reform, redistributive measures, institutional change, and structural transformations. A development could be promoted, it was hoped, that could steer clear of politics and 'socialisms'.[1]

In order to create conditions in which widespread adoption of the new technology could be expected to take place, programmes had to include the following components:

1. A technological 'package' or recipe produced in scientific research centres and designed to fit the environmental conditions of the region in which it is to be applied;
2. Arrangements whereby knowledge of this technology could be communicated to cultivators;
3. Measures to ensure the availability of physical inputs, i.e. HYV seeds, fertilizers, pesticides, machinery and fuel;
4. Measures to favour the prospect of profitable sale sufficiently attractive to compensate for the greatly increased production costs and risks involved;
5. Indispensably, some system of credit so that the payment for inputs and additional cultivation expenses could be financed, pending the receipt of income from the sale of the product after harvest.

---

[1] This point of view is well expressed by Hopper in his contribution to the Rockefeller Foundation, *Strategy for the Conquest of Hunger*, in an article entitled 'Investment in Agriculture: the Essentials for Pay-off':

Let me begin my examination of the essentials for pay-off by focusing on public policy for agricultural growth. The confusion of goals that has characterized purposive activity for agricultural development in the past cannot persist if hunger is to be overcome. National governments must clearly separate the goal of growth from the goals of social development and political participation. . . . These goals are not necessarily incompatible, but their joint pursuit in unitary action programs is incompatible with development of an effective strategy for abundance. To conquer hunger is a large task. To ensure social equity and opportunity is another large task. Each aim must be held separately and pursued by separate action. Where there are complementarities they should be exploited. But conflict in program content must be solved quickly at the political level with a full recognition that if the pursuit of production is made subordinate to these aims, the dismal record of the past will not be altered.

I. INDIA

Since Independence, officials in India have stressed the importance of agricultural development and the government has
tried out a variety of food production programmes. The
Fourth Five Year Plan (1969–1974) lays down two main
objectives in this field: to provide conditions for a 5 per cent
annual production increase over the following decade; and to
enable the largest possible number of rural people to participate in the benefits of increased productivity.

The means of achieving rapid but permanent improvement
in the agricultural sector, and even that sector's relative
priority, have been subject to diametrically opposed views,
which have been reflected at various times in different
programmes. The first approach applied after Independence
was 'extensive': the available resources and technology were
dispersed over the widest possible area, in the hope of producing the fairest possible spread of benefits. It was exemplified by the community development programme, which
emphasized the integrated development of community life
rather than increased agricultural production as such.

Fifteen community development projects, each covering
about 100 villages, were started in 1952, with Ford Foundation financial assistance. The idea was to mobilize rural
dwellers for labour-intensive agricultural productivity projects,
supported by certain land reforms, by new village co-operatives
in which the State would be a partner, and by national extension services. But the supporting reforms and institutional
changes were not enough to provide the bulk of the rural
population with the access to credit and the agricultural
inputs they needed for intensified production.

In 1959, a team of Ford Foundation agricultural experts
was called in. Their recommendations for a selective and
intensive approach among farmers and among districts led to
the winding down of the extensive programmes and the initi-
ation of the Intensive Agricultural Development Programme
(IADP) in 1960–1. The programme was launched in the best
agricultural district in each State. It featured a package, thereby giving India several years of experience in this approach
before the advent of HYV packages.

The IADP package included better seeds and implements, a balanced dose of fertilizers and pesticides, and recommendations about proper soil and water management. It also envisaged a complex of services: technical staff, availability of credit and inputs, land-and-water improvement, storage, maketing, and mechanisms to ensure reliable prices for the cultivators' produce. A new working link was built between research and extension services by the building of a number of agricultural universities, supported by the State Governments and often given assistance by US educational institutions. These have played a significant role in evolving HYVs and accompanying technological practices. Another important feature was a built-in process of assessment and evaluation—an Expert Committee to guide a bench-mark study and subsequent operational and analytical studies (*D. K. Desai, 1969*).

One-Crop Concentration

Policies were adjusted to utilize and promote the new high-yielding foodgrains in the mid-1960s, but still on the basis of the Ford Foundation recommendations. The programme became known as the New Agricultural Strategy (NAS).

As set forth in the Fourth Plan Draft, the NAS was an attempt to consolidate and intensify the old and new elements in the food production programme. It concentrated on one-tenth of the cultivable land and to a great extent on one crop: wheat. Necessary inputs were to be made available even at the expense of scarce foreign exchange. Domestic and foreign investments in the manufacture of fertilizer and other inputs were actively encouraged. Research and extension were reorganized and coordinated. The distribution of rural credit was extended. As the programme developed, increasing government attention was given to price and procurement policies (see *Infrastructure* below).

1. *Administration*. The NAS reorganized the over-all administration of rural and agricultural development. One secretary was put in charge of a combined Department of Agriculture, Co-operation and Community Development, with agricultural research the separate responsibility of the Director-General of

the Indian Council of Agricultural Research (ICAR). There are ten branches of the Department: Production (crops and animal husbandry); Land and Water; Soil and Water; Inputs; Credit and Marketing; Machinery; Fisheries; Forests; Land Reforms; and Agricultural Census and Administration Co-ordination. There are also four specialized offices: the ICAR; Directorate of Economics and Statistics; Directorate of Extension; and the Agricultural Prices Commission (*Krishna, 1971*).

The HYV programmes, though conceived on an India-wide basis, have been carried out to a great extent through the apparatus of the individual States, and vary somewhat according to the States' agricultural and administrative characteristics.

A summary follows of the programme in the State of Rajasthan:

Under the guidelines of the National Plan, complemented by further instructions from the central government, the State HYV programme is formulated and implemented by the State Directorate of Agriculture, through an informal committee made up of the Agricultural Production Secretary (chairman), State Bank Manager, State Co-operative Manager, Registrar of Co-operative Societies, and the Director of Agriculture himself. Technical specialists at all levels are also attached to the Directorate. There are direct links with both Regional and District agricultural officials. At District level the Agricultural Officer (DAO) has sole responsibility for planning and implementing the HYV programmes. District production plans are drawn up by an Agricultural Production Committee, of which the DAO is executive secretary and which includes several representatives of the District Council or panchayat (see below). The District co-operative marketing societies (for procuring and distributing fertilizers and pesticides) and the District central co-operative bank also take part in the Committee.

State officials at all levels are sent to HYV training courses dealing with the significance of the HYV programme, providing technical information and arranging for inputs.

The Department of Agriculture acquires seeds from the National Seed Corporation and the Agricultural Universities;

the seeds are multiplied at mechanized farms (30 in Rajasthan) run by the State Agricultural Directorate. HYV financial requirements are supplied by the Reserve Bank of India to the Rajasthan State Co-operative Bank against government guarantee. The State Co-operative Society buys fertilizers and plant protection chemicals on this account and distributes them to agents and to co-operative marketing societies. They are then sold to individual farmers for cash or to co-operative members as crop loans.

In all States there is an administrative unit covering up to 100 villages called a Block, and in almost all States operational authority for the food production programme at that level is vested in one officer, the Block Development Officer (BDO). The Block as an administrative unit was devised under the earlier community development programme; the BDO has retained a key role in the execution of the rural development programmes. He co-ordinates extension officers from all relevant departments (e.g. agriculture, health, education, co-operatives) and supervises the Village Level Workers (VLWs) who serve as agents linking the cultivators themselves with the community development and HYV programmes.

Each VLW has six to eight villages that he advises on HYV and development projects. Investigators have found direct correlation between the degree of acceptance of the VLW by villagers and the village's rate of adoption of the new technology (*Danda and Danda, 1971*).

2. *Infrastructure*. While drawing on many of the arrangements set up under the earlier rural programmes, the government's New Agricultural Strategy has focused on the infrastructural supports particularly required by the HYV programme: price incentives to farmers and procurement and distribution of their produce; credit to meet the greater expenses of the new technology; improved and expanded irrigation; adequate and timely distribution of inputs, etc.

Regulation of the procurement, distribution, and pricing of foodgrain has been a continuous concern of the government, which has tried to balance the consumer's welfare against the provision to cultivators of increased incentive as well as the promotion of fairly uniform prices among regions and at all seasons. It has also been a controversial and sensitive subject, affecting the interest of all segments of the population.

Until the advent of the NAS, central control of foodgrain marketing tended to expand and contract according to current trends in domestic production and the availability of imports. On the whole, prices were kept as low as possible for the consumer, by control over procurement prices where possible, by subsidies on the consumer price, and by liberal import policies (*Ray, 1970*).[2]

But as the NAS began to be put into effect, the government sought the farmers' co-operation in increasing their cereal production by placing a new emphasis on attractive prices for their produce. Two central bodies were established in 1965. The Food Corporation of India (FCI) became the agency for procurement, import, distribution, storage, and sale of foodgrain. It was to purchase grain in government-regulated markets where these were in operation, in local markets or FCI collection centres, and to sell to State governments, District Collectors, authorized dealers or fair price shops (usually privately-run shops with a concession to deal in the government-subsidized foodgrains).[3]

At the same time the Agricultural Prices Commission (APC) was formed from economic experts to determine minimum support prices for foodgrains. The prices were to be announced in advance of sowing as a form of guarantee to encourage maximum production effort. The APC findings have represented only technical advice to the States' Chief Ministers, who have decided the final price by consensus at the Chief Ministers' Conference. These final support prices

[2] For example, in the early 1950s the government took over all inter-State grain transfers in an attempt to arrest prevailing high prices. Meanwhile the area under cereals was increased, leading to a rise in production and a sharp fall in prices. In 1955, the government instituted support prices for wheat and three coarse grains and again permitted free movement of grains between States, except in certain metropolitan areas. As production fluctuated, short-term measures were adopted. But when production declined radically from 1963–4, State zoning (i.e. restriction on grain movements) was reintroduced and by 1965 foodgrains were compulsorily procured and were rationed in the large cities. Great quantities of wheat were imported and subsidized to keep prices low.

[3] Information kindly provided by A. Menamkat from data gathered for his Ph.D. dissertation at the University of Fribourg, 'Development Problems and the Role of Credit Co-operatives in Indian Agriculture'.

have generally been higher than the APC recommendations (*Krishna, 1971*, p. 87).

To ensure that the poorer consumers should get part of their food at prices lower than prevailing market prices, the government has in the past compulsorily procured a portion (recently as much as one half) of traders' grain acquisitions, which has been distributed and sold through the fair price shops. However, procurement has not come up to expectations in many areas. The Food and Agricultural Ministry announced in January 1974 that in view of the unsatisfactory working of this procurement/distribution system, it no longer intended to expand its system of purchases, but would instead curtail it and remove controls on the free movement of the grains (*Economic and Political Weekly*, 2 Feb. 1974).

Increased Credit

For the supply of agricultural credit, the main institutional role had been assigned to the government co-operatives. The Fourth Plan assigned them an important, though not exclusive role.[4] Agribusiness was expected to facilitate hire-purchase of agricultural machinery and pumps, and special emphasis was placed on the (now mostly nationalized) commercial banks. For major, macro-level development projects, medium- and long-term credit was entrusted to the Agricultural Refinance Corporation, a consortium of commercial banks set up in 1963.

The flow of credit into the agricultural sector increased significantly after the nationalization of fourteen of the major commercial banks in 1969.[5] The commercial banks, however, tend to operate to the detriment of the co-operatives because, with their large network of branches, they can offer better remittance and banking facilities, thus taking support from the co-operatives. There have been a number of attempts

---

[4] The Revenue Department issues short-term *taccavi* loans, secured by mortgage on land and recoverable by the Revenue Department. In some States this type of loan is reserved mainly for such emergencies as crop failure, famine, and flood, and in others (e.g. Uttar Pradesh) up to 40 per cent of short-term official loans may be *taccavi* loans (*Hunter, 1970*).

[5] Kahlon and Singh's study for UNRISD (1973, II) reports in one Punjab district, Ferozepur, an increase in the level of commercial banks' agricultural financing from Rs. 912,500 in 1969–70 to Rs. 2,136,400 in 1971–2.

at co-ordinating the two sectors, but they have not yet
succeeded (*Krishna, 1971*).

Primary co-operative societies distribute most short-term
crop loans. The ultimate co-operative credit source is the
Reserve Bank of India, which channels credit through State
Banks or the State Finance Corporation to Co-operative
Unions or District Co-operative Societies and thence to the
primary societies. If a primary society is in default (less than
60 per cent repayment) no credit can be given to it the
following year, and it must make its own recovery from
individuals. Larger medium- or long-term loans for capital
improvements are usually given through a land mortgage
bank, land development bank or agricultural finance corporation.

The amount of short-term credit disbursed has increased
rapidly since the early 1960s, as has the share of credit
handled by the co-operatives (*Krishna, 1971* and *Hunter,
1970*), but the main sources of credit are still private—
relatives, friends, moneylenders, and landlords. The co-
operatives often have histories of inefficient or opportunistic
management. Although they were intended to provide the
cheaper credit needed to make the new technology profitable
for the majority of farmers with little or no surplus of their
own to invest in it, in some villages this credit has been
appropriated by large landlords of the dominant caste group,
who influence the local co-operative as well as the local
council. Tenants and owners with little security generally
must pay a great deal more for private credit, which is sometimes co-operative credit borrowed by the landlord and re-lent
(*Parthasarathy, 1973*).

Other government projects, notably irrigation, roads, rural
electrification programmes, and fertilizer plants, have been
set in motion on the national and State levels. Professor V. S.
Vyas, an Indian economist, identified the States which have
made significant progress in irrigation, road building, and
rural electrification as those that have taken significant
strides in agricultural production (*Vyas, 1973*).

Too Little Water

The amount of irrigated land increased by 17 per cent between

1960-1 and 1968-9, at which time the net irrigated area was 28.73 million hectares, or about 21 per cent of the net area sown (*Easter, 1974*). At Independence, India inherited an elaborate, commercially-oriented irrigation system, with spheres of responsibility overlapping between the Canal and Agriculture Departments. The public sector irrigation supply continues to operate on commercial principles: facilities tend to be sited where satisfactory returns on installation costs can be expected rather than where benefits would be most widespread. Administrative overlapping continues, and some operations still present many problems: State tubewell users have often complained of too little water, especially reductions without warning, poor servicing of machinery, and, most frequently, failure of the electricity supply (*Whitcombe, 1974*).

An All-India Rural Electrification Corporation was set up in 1969 to finance electrification programmes and to help Electricity Boards to invest their funds appropriately. Government directives to the Corporation included financing criteria of economic viability and increased agricultural production, but also instructed that preferential terms be given to relatively backward areas and that the Corporation promote a co-operative method of distributing rural electricity. Several pilot consumer co-operatives have been set up. Daya Krishna reports that the Corporation has been financially viable in the Punjab, though not in Kerala, but points out that the benefit to farmers and to the rural economy as a whole cannot be disputed (*Krishna, 1971*).

The rural electrification programme and institutional loans have made possible an 'almost explosive expansion' in the use of individually owned wells and pumpsets (*Krishna, 1971*). State tubewells have been controversial because they have a much higher capital cost than private ones and to recuperate this are operated much more intensively. This makes them deteriorate faster and break down more often, so that they then tend to become unreliable, unprofitable, and more expensive to the farmer.

In a study of Indian irrigation policy for UNRISD, Elizabeth Whitcombe (1974) reported that in tours of more than 600 Blocks, it was rare to find more than a third of the State tubewells in working order. At the same time, they tap a deeper source, and may therefore eventually justify their higher cost. Meanwhile, the private tubewells can mainly be utilized only by the larger farmers (*Mellor and Moorti, 1971*).

The Third Five-Year Plan provided for a dual-purpose Rural Works Programme, using surplus rural labour to build small-scale infrastructural projects. On the basis of the projects completed in the Third Plan period, Professor Vyas has estimated that Rs. 100 would generate 50 man-days of employment when invested in a rural roads programme, 59 in soil conservation and 94 man-days invested in minor irrigation (*Vyas, 1973*).

State governments involved in intensive HYV programmes have also established their own infrastructure agencies. For example, Punjab State, in addition to the Agricultural University mentioned above, also established the Punjab State Agro-Industries Corporation and reactivated the Punjab State Co-operative Supply and Marketing Federation. It also launched a crash programme of village road building in 1968, which produced almost 2,000 kilometres of link roads in 1969–70 alone.

3. *Community development and local councils.* Although no longer given priority, the community development programmes remain an important element in the Indian countryside. By the end of the Third Plan they had reached all parts of India: in April 1971, 4,893 community development Blocks covered all 566,900 Indian villages. The programmes have included cultivation projects, minor irrigation, animal husbandry, social education, village and small industries, health, rural sanitation, family planning, and communications.

The programmes have varied from State to State in emphasis and effectiveness but their results, though by no means what the planners had hoped for as regards viability or local finance for projects, have not been entirely negligible. Agricultural demonstrations organized by the Blocks increased seven-fold between 1967 and 1972. The amount of seeds distributed increased by 12 per cent and fertilizer by 64 per cent. Areas using plant protection measures increased by 95 per cent (*Singh, 1973*).

Perhaps the main contribution to the programmes thus far is to make the villagers aware of the need for and possibility of change and to bring them into greater administrative contact with the district, state, and nation. One Indian sociologist (*Bandyopadhyay, 1972*) believes that the main impact of the programmes has been as an agent promoting infrastructural improvement and incorporating the local society into the market system of metropolitan society. He describes community development villages in a formerly isolated district of West Bengal:

All-weather metalled roads across the district and a concrete bridge over the water . . . have linked these villages quite closely with the markets of the wider urban-industrial complex of Asansole, Durgapur, Burdwan and even the Calcutta area. The distance of Calcutta from Suri, the district headquarters, is about 150 miles and can be covered by car in five or six hours. Buses nowadays connect almost all the spots of some importance. Middle-men and agents of wholesalers from urban markets are now regularly visiting the villages in large number to purchase wheat, paddy, pulses, milk, eggs and vegetables (especially cauliflowers, cabbages, potatoes and onions). (pp. 7–8)

In 1957, an Indian study team suggested that the community development movement was ineffective in generating local support because it was administered by government authorities and depended on government initiative and funds. The team recommended transferring a certain amount of power and responsibility for community programmes to elected bodies at village, block, and district levels. The *panchayat raj* or elective council system was the result of this recommendation.

By March 1971, village *panchayati* covered 98 per cent of the rural population. They are expected to deal with sanitation, crop experiments, promotion of agricultural and cottage industries, birth and death registration, and other functions delegated by the States. Block and district level *panchayati*, made up of the presidents of the next lower stages, supervise the latter and have specific responsibility for primary education, health, sanitation, and communications. They are all authorized to levy certain taxes to cover their projects, but their incomes have been inadequate, despite grants from State government and educational institutions.

Though leadership cannot be directly hereditary as in traditional village government, since the *panchayati* are elective, domination of them by landlords and top castes has often continued (*Bandyopadhyay,.1972* and *Beteille, 1965*).

4. *HYV projects*. In addition to the special schemes for small and disadvantaged farmers described in Chapter X, there are a number of ongoing schemes specifically to promote HYVs. One is the centrally sponsored 'Farmers' Training and Educational Scheme', begun in 1966–7. It was to cover 100 HYV districts by the end of the Fourth Plan and provided for 15 demonstrations in each district under a team of experts in soil, agronomy, plant protection, and agricultural engineering. The three main components of the scheme were (1) farmers' training, including demonstrations, special courses, discussion groups, and conducted tours; (2) farm broadcasts; and (3) organization of functional literacy courses.

Two successful rice schemes have been reported from Tamil Nadu (*Chinnappa, 1972*). The Navari Crash Programme has been a unique and apparently successful means of encouraging farmers to adopt recommended practices by arranging, at a very small cost to the cultivator, the delivery to his farm, by van, of the inputs needed for the planting of IR-8 paddy.

Another Tamil Nadu programme is the Multiple Cropping Scheme, introduced in 1967–8 as part of the New Agricultural Strategy. Crop rotation of HYV rice with jute, pulses or millet was instituted, irrigation facilities were being improved, and there was continuous soil testing, the necessary fertilizer being supplied when the soil appeared exhausted by repeated cropping. The area covered by the scheme increased from 100,000 hectares to 530,000 hectares from 1967–8 to 1971–2.

Privilege Maintained

The picture that emerges from a review of rural and agricultural developmental programmes is a confusing and less than encouraging one. Few countries have made such serious efforts on so large a scale to bring about a transformation of traditional agriculture. Yet in spite of some notable successes, the current situation remains critical, both as regards over-all production needs and as regards participation in agricultural production by the rural masses as an effective means of achieving and improving livelihood. Though we do not feel called upon to venture an over-all judgement of the situation, a number of points have been made by observers with sufficient insistence to be worth noting here.

In view of the inherent weakness of the administrative system, the coherent interplay of the many district programmes becomes extremely difficult. A further set of obstacles in the way of large-scale planning for rural development is to be found in the degree of autonomy exercised by the States, which among other things retained power for agrarian legislation and almost universally resisted central pressures for a redistribution of resources in land and water.

At the village level, the apparently sound principle of devolution of responsibilities put control of government resources and services for development into the hands of higher-caste land-monopolizing élites, thus ensuring the preservation and expansion of privilege rather than local transformations.

II. INDONESIA

The case of Indonesia is especially interesting, since a number of different programmes aimed at raising the level of peasant technology have followed one another, partly in response to changing political and economic conjuncture and partly in the logic of learning by experience. The first of such programmes (1960–3) was based on a territorial principle, with the operation of a 'paddy centre' for selected areas of approximately 1,000 hectares each. The centre provided extension seeds, fertilizer, and credit but the product was compulsorily procured at a price that fell below that of the market. To this disincentive was added dissatisfaction over the credit terms and an inefficient staff.

The Mass Guidance Programme (BIMAS) began in 1963 and followed on from the earlier Plan, which was deemed to have failed in its objectives. The new programme was tried out with all peasants in selected areas, to whom fertilizer, improved seeds, and insecticides were distributed on credit, to be paid in produce after the harvest. Distribution was accompanied by 'intensive grass-roots guidance' by politically motivated students. An attempt was made to give flexibility to the programme in spite of its compulsory character by insisting on a two-way information flow.

The political *volte-face* of 1965, with the massacres of activists that followed it, penetrated deeply into the main rice-producing areas. The programme was continued and the extension crops expanded, causing a further dilution of its quality. The results of the programme in terms of productivity did not come up to expectations. Political strife weakened institutional co-operation, and farmers were reluctant to continue their participation.

In 1967, a decision was taken to divide the programme into two layers—one of these being a continuation of BIMAS for the more traditional cultivators, and the other catering for cultivators disposed to adopt more intensive cultivation, and having less encumbered access to resources. This change in policy coincided with the introduction of IRRI seeds, and a move away from a politically motivated movement to a more severely technocratic one:

With the idea now gone . . . that agricultural development was a social and political as well as an economic transformation, the government could devote its attention to specific factors of production. . . . Gone are the discussions of nationalistic fervour, gone are the allusions to the role of students as a catalyzing force, gone is the emphasis even on local extension work, gone are the references to forms of 'exploitation'. Enter a new generation of propositions concerned with a more limited set of problems: can the government get the seeds to the farmers on time for planting? Are the fertilizer recommendations 'optimal'? Are the new seeds sufficiently resistant to local disease? . . . In short, would the planners who now had taken the (rice intensification) programme from its recipient in connection with peasant political movements, be able to get the right physical inputs out to the right places in the right amounts at the right time? (*Franke, 1972*, pp. 32–3.)

The BIMAS Programme that followed seems to have been an attempt to impose a technically efficient system of production willy-nilly upon cultivators. Sophisticated know-how from the Ford Foundation, USAID, and IRRI was brought in by the military government, and the organization of a large part of the programme was contracted out to large commercial agribusiness corporations,[6] the largest commitment being undertaken by one of the most modern producers of chemicals—CIBA, the Swiss-based transnational company. By the end of 1969, opposition to this programme was widespread amongst the more enlightened officials and intellectuals, and there was active resistance by the peasantry.

The programme as it was planned had the defect of treating the cultivator as an object. He was drafted into the programme without choice, delivered a standard 'package' without consideration of local soil/climate specifications and given a subsistence loan in cash. His crops were indiscriminately submitted to aerial spraying of pesticides and a share of his crop was requisitioned, equal to one sixth of a standard anticipated harvest, in repayment of the loan he had not requested.

Such were the formal conditions. But complaints from cultivators revealed that supplies were frequently 'diverted', that the subsistence loan was not always paid, that the aerial spraying killed animals and fish, and that the one sixth crop repayment frequently bore no just relation to what had been received.[7] However, one advantage enjoyed by this programme in comparison with its predecessors was the reduced economic uncertainty provided by this form of repayment when it was carried out correctly.

Improved BIMAS was introduced in 1970/1 with a pilot project in the Jogjakarta area. Its provisions were numerous. Village banks channelling credit directly used land as collateral while sharecroppers and tenants were allowed to choose the amount of credit they required and the form of repayment, whether cash or kind; the bank was to announce the rice purchasing price before the planting season so that farmers could include it in their calculations; guidance was restricted to extension services; village storage sheds were established

---

[6] Franke reports that one such corporation was owned by a group of enterprising generals!

[7] In a personal communication (31.10.78) Wertheim pointed out that the government incurred a heavy debt to CIBA, which it attempted to recuperate from the peasantry.

where cultivators could leave their rice at harvest time on a credit basis, so that sales on the market could be made propitiously and in a leisurely manner.

The management of the banks and storage sheds, together with the extension service, formed a village BIMAS executive board under a single administrator, supported by government credit. Eventually, the board was to become a co-operative.

Since there is no land reform, those without land are still dependent on the bank administrator for credit—and the latter are still likely to prefer land as collateral.

In some areas cultivators have complained that they could choose only among a few complete input packages. It is also said that the packages are often broken into—the cash allowance is mainly taken, but also fertilizer, though this is less regretted by cultivators, who prefer free market chemicals. The packages also often arrive too late for optimum use, which has increased cultivators' reluctance to repay. Credit procedures are unclear, time-consuming, and cumbersome. Moreover, the credit administrator rarely grants as much credit as the cultivator considers necessary. The 1971 Agro-Economic Survey showed that fixed-rent tenants received the most credit per hectare, followed by owner-operators. Share tenants did worst. The highest repayment successes were shown to be in areas with good irrigation, fertile soil, dense population, very small farms, and rice monoculture.

Clearly, some lessons had been learnt from experience.

Sajogyo (1973) reports that for the wet season of 1971/2, 264 'village-unit' organizations, each servicing a cluster of villages with an average of 677 participating cultivators, covered 170,000 hectares. There was one Bank representative for every 462 farmers and one extension office for every 512 farmers. In each cluster four kiosks were set up to sell fertilizer and pesticides, each one serving 187 cultivators.

Sajogyo criticizes the passivity of the relations between peasants and agencies, and he notes the difficulty of persuading farmers to form their own self-help organizations to 'support, meet and match the services of government agencies'. He says that the agencies still have no clear operational concepts for the development of enduring cultivators' projects.

Village headmen are selected to represent cultivators'

interests but they are more inclined to turn into representatives of the government than of their constituents. So the cultivators (he says) have no say in the local planning of the BIMAS programmes, even though these vitally affect their interests.

III. SRI LANKA

Sri Lanka's food production programmes are based on improved local rice varieties rather than on those tracing their ancestry to IRRI. The framework of the programmes is the Food Drive, a major campaign begun in 1967 with the main aim of reducing food imports, which still take almost half of the country's foreign exchange earnings. It was hoped to overcome the external payments crisis and make foreign currency available for the capital goods and raw materials needed in the industrialization process. A great deal of responsibility has thus been laid on the food production sector for industrial development and consequently for improving the country's deteriorating employment situation.

Sri Lankan governments have an even more direct reliance on the food sector: almost since Independence in 1948, food consumption subsidies—mainly for rice—have been considered a political necessity by each successive government.

Under the colonial regime, the country and its economy were dominated by plantations of tea, rubber, and coconuts, and most of its food was imported, including a great deal of its staple rice. Domestic rice was raised on relatively small farms, often with poor irrigation and drainage. After Independence, political and economic change put an end to an essentially bi-modal agrarian structure and peasant agriculture was given central place in the economy as the main source of domestic food. A policy of intensive agricultural research has been pursued which, by concentrating on improving indigenous rice varieties and accompanying practices, has produced for Sri Lanka an uniquely appropriate technological base and made unnecessary any dependence on foreign HYV seeds.

The Food Drive started with three basic assumptions: food production, especially rice, must have absolute priority, the agricultural infrastructure must be made adequate, and farmers must have price incentives to produce more. The importance of the Drive was underlined by the fact that the Prime Minister took direct charge of it. Policy was the responsibility of a special Cabinet sub-committee, whose decisions had the status of Cabinet decisions.

In addition to the technological foundation described above, the government has been able to use the framework provided by earlier programmes and legislation stretching back as far as thirty years in some cases: a Guaranteed Price Scheme for rice, institutions for credit and self-help, attempts at certain tenancy reforms, and a consistent land settlement policy.

The Paddy Lands Act was intended to obtain security of tenure for tenant rice farmers by making tenancy inheritable, and to set maximum rent for tenants. Because the machinery set up to implement these provisions is vulnerable to manipulation, the Act's aims have not been satisfactorily realized. And where the Act has been effectively implemented, however, it is reported to have caused a break in the patron–client relationship between tenant and landlord that has left the tenant with less security of tenure and less access to capital, in the absence of any co-ordinated government action to replace traditional landlord functions (*Hameed, 1977*).

Cultivation Committees, each made up of twelve members elected by the farmers of a village, were created primarily to promote implementation of the Paddy Lands Act by maintaining a register of land ownership and tenure. But other functions assigned to them are particularly important for increasing production and spreading new technology: guaranteeing loans, procuring and distributing inputs, and collaborating with extension officials.

The Committees have had some success, especially with carrying out minor irrigation works under government contracts, but because of difficulties in collecting the acreage levy that supports their functioning and in some cases because local élites achieved control over them,[8] most of their

---

[8] The Committees are 'invariably dominated by the fairly powerful landed class who manipulate affairs to the disadvantage of poor peasant cultivators'

activities have lapsed or, as in the cases of land registers falsified to conceal illegal tenancy, have been subverted.

Hameed says that a major obstacle to increased productivity is inadequate irrigation and drainage. A number of major and minor irrigation schemes are under way, but they have focussed more on bringing new lands under cultivation than on improving existing lands. There has been little government effort directly to help farmers or village bodies to obtain better local irrigation facilities, except for that given by the Cultivation Committees. Where the Cultivation Committees are ineffective, or where there is a lack of local funds, the irrigation is often poor.

In an attempt to overcome another primary problem—the farmers' inadequate resources for the new technology—the Food Drive programme includes a New Agricultural Credit Scheme. The Scheme has more than doubled the amount of money lent to individual farmers. One aspect that has facilitated the widespread adoption of HYVs and new agricultural practices is the crop loan system. Loans partly of inputs and partly of cash are distributed to a maximum amount of Rs. 370 per HYV acre. Only Rs. 262 is allowed per acre under traditional varieties. However, amounts lent under this scheme have declined greatly since 1969, for reasons set forth below.

Agricultural credit is extended through multi-purpose co-operatives, which were organized in their present form in 1957. There are about 4,500 of these societies, which also distribute inputs and procure surplus paddy at the village level. But the unsatisfactory working of the co-operative credit system has greatly prejudiced the over-all role the co-operatives are meant to play in the agricultural sector and particularly in the Food Drive. Their lending methods are 'unduly complex, time-consuming, inflexible and costly' (*Hameed, 1977*). Defaulting is common and many farmers have come to look upon government loans as outright grants,

(Selvanayagam, in *Hameed, 1977*). In Minipe, where there is less land inequality, the Committees are fairly successful (*Hameed, 1977*) but in Palannoruwa (Selvadurai, in *Hameed, 1977*) the Cultivation Committee had to stop functioning in 1970 because it could not collect all of the levy and some officers had embezzled funds.

or at least give priority to paying off loans to the village moneylender. Co-operatives make no distinction between deliberate defaulters and those whose poor harvests prevent repayment. So, despite a remarkable growth in the number of institutional agencies, it is friends, professional money-lenders, landlords, traders, and so on who continue to be the principal sources of rural credit.

A crucial facet of the Food Drive has been the provision of price incentives to increase production. A Guaranteed Price Scheme introduced in 1948 was subsequently turned over to the multi-purpose co-operatives. The Scheme has maintained a reassuring floor price for rice and to some extent seems to have encouraged farmers to produce. But it too has suffered from the defects of the co-operative system: defaulters cannot take advantage of the government price, and there are inconveniences, delays, and irregularities in reception and payments.[9] A Monopoly Purchasing Act gives the Government Paddy Marketing Board the exclusive right to buy and process paddy.

Meanwhile, the government's efforts to provide price incentives have been helped by outside circumstances. The rise in world rice prices in the second half of the 1960s not only lent urgency to the Food Drive but at the same time in itself provided one of the most important production incentives by its effect in raising domestic rice prices.

In an attempt to overcome farmers' reluctance to take on the added risks of the new seeds, a Crop Insurance Scheme was introduced in 1958/59. It is intended to indemnify up to 50 per cent of crop value against lack of adequate water, floods, diseases, and crop damage from wild elephants, etc. It is not working well; in fact, the indemnities paid exceeded the premiums collected by more than Rs. 5 million in the first ten years. One obvious problem is that this scheme too is implemented at village level by the Cultivation Committees and co-operatives. Another is that extension workers do not explain the scheme to farmers so that they can see the advantages (Hameed, 1977).

---

[9] Selvadurai (in Hameed, 1977) reports that in Palannoruwa a co-operative manager was also a private trader, buying privately what the co-operative rejected.

The neglect of the extension services as a major instrument of development policy is also one of the main reasons for the general use of less than the optimal amounts of fertilizer and pesticides. The government subsidizes fertilizer by 50 per cent. Importation and wholesale distribution of fertilizer and other HYV chemicals is carried out by the public-sector Fertilizer Corporation. But the inadequacy of institutional agencies such as the extension services and the co-operatives (which handle much of the retail distribution) and of such infrastructural facilities as irrigation works (since fertilizers are relatively ineffective where water is inadequate) have partly nullified the subsidy and other government efforts to promote fertilizer use.

The government has incorporated the technical elements necessary for widespread HYV use into its long-term research programme for seeds and cultivation practices. But here again, the elaborate and costly state system of producing HYV seeds has in some localities been prejudiced by poor control on the part of the public authorities, which have sometimes distributed poor quality seeds to the multi-purpose co-operatives. Two innovations have generally proved popular and effective, however. A system of 'mini-kits' was introduced in 1971 to help farmers to test and produce their own seed. Each farmer was given free five or six new varieties, each enough to cover a five-foot strip, along with the recommended fertilizer and chemical dosages. The following year ' production kits' containing a pound of a new seed variety plus the other required inputs were distributed for a nominal price.

Despite their shortcomings, the measures to increase food production taken by post-independence governments in Sri Lanka, culminating in the Food Drive, more than doubled domestic rice production between 1950 and 1970, and increased yields by over 65 per cent. But unrest in the countryside has become more and more visible, particularly since 1970. There has been a substantial decrease in rice production and the use of modern techniques, to such an extent that imports had to be increased in 1972 and again in 1973. The Food Drive, with its limited though urgent aim of self-sufficiency, has done little to reduce rural unemployment fed by a high birth rate, or to cater for the growing proportion of educated rural youth. Where production increased, cultivators' real income did not increase proportionately since input prices also increased.

Because the required working capital could not find its way to the hands of the numerous small cultivators, they responded least to and gained least from the new technology. The current Five-Year Plan sets up District Development Councils; these, along with a reorganized primary co-operative network and rural banks, are meant to provide finance and other facilities to small cultivators, tenants, and landless labourers. But the Plan's proposals have not yet been effectively implemented.

IV. MALAYSIA

Malaysia has also utilized HYVs as an important development weapon, but no new projects or programmes have been designed for them alone. Finance, skilled personnel, and other scarce resources have been devoted to more basic programmes, such as improving and extending irrigation facilities, establishing farmers' associations, making institutional innovations in the input and output marketing structure, and intensifying and reorganizing research activities.

The objectives of the Malaysian Government's policies for the rice sector have been to increase the incomes of paddy cultivators, to reach 80 to 90 per cent self-sufficiency in rice and to supply rice to consumers at a reasonable price, within the framework of Malaysia's 'New Economic Policy'.[10] Rural incomes are much lower than urban ones, paddy farmers have among the lowest rural incomes and even among paddy farmers there are great regional inequalities. At the same time, rice is the staple food of all Malaysians; any rise in its price can be a hardship for the majority of consumers. Finally, Malaysia is not unique in preferring not to depend on other countries for its staple food, or not to go on bearing a heavy food import bill.

---

[10] For a discussion of the 'New Economic Policy' in this context, see Bhati (1976).

The public sector has been called upon increasingly to implement rice production policies. A National Paddy and Rice Authority (NAPRA) was established in 1971 to co-ordinate production, processing and marketing and is responsible for over-all national policies for the rice industry, including the setting of prices for buying and selling. This body must balance considerations of production incentives and national self-sufficiency against immediate consumer well-being.

As in India and some other developing and developed countries, policy execution is often thwarted or at least hindered by the division of powers between Federal and State organs. NAPRA is responsible for co-ordinating rice production policies at the national level between, for example, the Ministry of Agriculture and the Ministry of National and Rural Development. At the State level,[11] a Department of Agriculture receives directives from the State Government and from the Federal Ministry of Agriculture. The lowest rung of the vertical structure leading to the Federal Ministry of Agriculture is occupied by the Junior Agricultural Assistant (JAA), attached to the farmers' associations (see below) where they exist and otherwise directly to the State Agricultural Department. The JAA does extension work and a large variety of other development tasks at grass-roots level. In practice, JAAs with sufficient training are difficult to find, and then are often given inadequate resources and inadequate rewards and prospects. The frequent ineffectiveness of this important sector has adversely affected extension services and HYV seed distribution in particular.

To promote the farmers' direct participation in increasing their production and income, legislation in 1967 provided for farmers' associations (FAs) modelled on those of Taiwan. They are meant to be grass-roots agencies for implementing agricultural development programmes and for transmitting farmers' needs and views to the government. The FAs each cover a number of villages and have an average membership of about 2,000. At village level, Small Agricultural Units

---

[11] This description is provided by the Penang State Department of Agriculture; others may differ slightly.

select representatives to the FA Board of Directors. An Act passed in 1972 gives the Minister of Agriculture the right to appoint a third of the Board members. The Act was the result of pressure from local politicians fearing loss of their control over local events. There was an unsuccessful attempt that same year to enact a law permitting any landowner to join a FA, instead of requiring that he get at least 50 per cent of his income from farming. This legislation would of course have diminished the voice of the small cultivators in FA affairs. In this way the FAs, though organizationally separate, are functionally fused into the State Agriculture Department's Extension and Farmers' Association Division.

Government programmes since the 1960s have been primarily directed towards large-scale irrigation and consequent new double-cropping opportunities, as the major step towards self-sufficiency in rice. The most ambitious project is the Muda Irrigation Scheme, which began operations in 1969/70. The Scheme covers about a quarter of the total paddy area in West Malaysia and has already almost doubled the amount of irrigated land that can be double-cropped. It is eventually expected to provide almost a third again as much paddy as is now produced in Malaysia.

At first an ineffective attempt was made to centralize authority for the entire scheme in the Ministry for Rural Development, with the Deputy Prime Minister at its head and with the Prime Minister and entire Cabinet leading an associated National Rural Development Council. The idea was to avoid overlapping and rivalries amongst States, Ministries, and political parties, but in fact the high-powered political make-up of the Council inhibited the efficiency of Scheme administrators. A Muda Agricultural Development Authority (MADA) and a similar co-ordinating authority for the Kemubu Scheme (see below) were accordingly created in 1970. Though ostensibly autonomous, with wide powers, they must have permission from the Ministry of Rural Development to secure loans. In practice, this has given the Ministry the control it needs to plan and implement integrated projects in a way the Council could not.

The other large project, the Kemubu Scheme, was begun in 1969 and is converting 19,000 hectares to double-cropping.

There has been ample evidence that the two schemes will do away with the need for rice imports. The government therefore decided in 1970 to bring no new land under paddy, though continuing to extend irrigation on existing land.

The Schemes have not been trouble free. An unexpected US$21.2 million worth of work in drainage and canal-building remained to be done after the foreign construction company had fulfilled its contractual obligations for the Muda Scheme. Malaysian engineers had to be trained on the job. There are also expensive, time-consuming problems of equitable water distribution to small farms on uneven slopes. Moreover, according to Randolph Barker of IRRI (1971), the Scheme has poor water control and is too inflexible to be used for any crop except rice.

The government has had a Guaranteed Minimum Price programme since 1949 implemented through private and para-governmental agencies, and largely financed by manipulating the price to consumers of imported rice, in this way maintaining incentives for local producers without inconvenience to consumers of local rice. A Federal Agricultural Marketing Authority was set up in 1965. It collects and disseminates marketing information, licenses paddy buyers, and generally regulates and tries to improve marketing facilities. Since 1973, NAPRA itself, in association with State organs, has taken over the wholesale distribution of rice and paddy, on the grounds that private dealers were not functioning responsibly and efficiently.

The importance of developing and applying new technology has long been recognized in Malaysia. There are government agencies for research and education, but they have lacked trained technicians and research facilities. Writing in 1973, Bhati reported that, while soils and water levels differ widely in Malaysia, and while the relative prices of various fertilizers had changed since IR-5 rice was introduced in 1968, recommended quantities of fertilizers and the recommended composition of the basal fertilizer mixture had remained uniform and static since then.

Other aspects of the government's rice programmes are free pesticides, the 30 per cent fertilizer subsidies for single-cropping areas, and certain land reform measures like the

Paddy Cultivators (Control of Rent and Security of Tenure) Act 1967. These measures, however, have not been effectively implemented.

The programmes have achieved some of their announced objectives. The goal of 80–90 per cent self-sufficiency in rice has since been reached, along with the objective of keeping the consumer rice price low. The objective of raising rice farmers' incomes has probably been achieved only to a very small degree and in the best double-cropping areas. Existing geographical inequalities of incomes have been aggravated by subsidizing the farmers' irrigation in new and highly productive areas.

To make these schemes pay, Malaysia's limited extension services and agricultural credit are still concentrated on the new double-cropping paddy areas. The government has recommended that farmers in other areas should grow other crops in rotation with paddy, or be assisted to take up other occupations. But it had not been able to back up its recommendations with the necessary resources at the time of writing.

VI THE ECONOMICS OF FARM SIZE

It is shown that the economies of scale enjoyed by big users of the new technology are not so great but that the big user nonetheless draws greater economic advantages from it than the small cultivator does.

Discussions of farm size in relation to the new technology are concerned with several different issues. Promoters of the Green Revolution want to know why small cultivators are often loath to take up the recommended package, or why they water it down, and there are many studies seeking to identify attributes and characteristics of small cultivators and to show that these are 'obstacles to modernization'. Unfortunately, these studies contribute little to our undertaking because of their common failure to penetrate the complexity of this technological transformation and the oversimplifications implied by concepts such as 'innovative behaviour' and 'adoption' as used in their field studies.

Protagonists of the Green Revolution strategy are also much concerned with second- and third-generation problems, especially the long-term consequences for agrarian structure. They are particularly anxious to defend the strategy, or perhaps to adjust it, in response to assertions that it aggravates socio-economic polarization. It is in this connection that prominence has been given to the scale-neutral character of chemicals and seeds.

It should be quite clear that the question of size of holding opens up major political issues for the obvious reason that as soon as one talks about people with large farms, people with small farms, and people with no farms at all in an agrarian society, one is talking about rich people, poor people, and very poor people. And when tenancy is discussed, the whole class structure of village social relations involving super- and sub-ordination, élite rule and patron-client dependence is brought into question.

Moreover, most of the discussion of farm size has been carried on within a conceptual framework invented to fit the operations of firms within a capitalist economic structure rather than livelihood support systems in pre-capitalist modes of production.

The evidence about size of holding[1] in relation to productive capacity as indicated by yields, and to the economics of profitability, will now be considered.

I. YIELDS AND SIZE OF HOLDING

The first point to establish is that very significant advances in yields have been achieved by the use of high-yielding varieties. This is shown in Table 8 for rice and wheat, with a single case of maize. It must, of course, be borne in mind that the localities for which this table gives yield data were, most of them, selected for the introduction of the technology on account of the favourable conditions found in them.

In spite of this, it was found that yields varied greatly from place to place and within each place there was great variation between farms sharing common climatic conditions. Some of this variation is widely reported to have been due to variation in the quality of seeds, while there were numerous cases of difficulties in dealing with the unexpectedly luxuriant growth of weeds due to increased fertilizer use and the occurrence of new pests and diseases, which proliferated on account of the density of growth of the new varieties.

But the three most important factors ensuring high yields seem to have been:

(i)   the adequacy and timeliness of moisture, which translates itself into the adequacy and effective control of irrigation;

(ii) the amount of fertilizer used; and

(iii) the quality of husbandry.

We must now ask how the yield-performance of small and large cultivators compared.

The availability and controllability of irrigation water has

---

[1] The evidence about size comes from the UNRISD studies and particularly from Dasgupta (1977), to whose works reference should be made for more detailed treatment.

undoubtedly been greater for large cultivators, the extent of whose lands made it an economic proposition to invest in one or more tubewells. With these installed, ground water was available when required, making farms independent of rainfall and weather-dependent canal systems. But as a single tubewell is an item of machinery that requires a large farm for its economic use, small farmers were only in exceptional cases able to enjoy this optimum system of irrigation. Thus, one factor favouring the larger cultivator over the smaller has been the spread of tubewells, associated with the great forward leap in wheat production witnessed in North-West Mexico and the Punjab (of both India and Pakistan).

In regard to the use of fertilizer and the quality of husbandry, the situation is more complex. But one important change in the yield-relationship between large and small farms seems to be taking place. Under conditions of the received local technology in Asia, the highest yields were commonly achieved by cultivators of the smallest plots of land. This was directly due to the very intensive use of family labour that could be applied to the tasks of husbandry, such as weeding and careful water management, without adding to production costs.

In areas where fertilizer use had already been promoted, it was not infrequent to find that the same smallholders were intense users of organic or manufactured fertilizer. In contrast, larger farms using wage labour achieved lower yields. Their non-family labour was less conscientious and had to be paid wages, and within this less intensive context, heavy investment in bought fertilizer was not undertaken.

However, several of the studies in optimum wheat-growing areas demonstrated that with the advent of the new technology, many large farmers in particular became heavy investors of operating capital both in increased labour time, for a more careful and complex husbandry, as well as in fertilizer, and ever more frequently they came to surpass the smallholders in intensity, in production costs, and in yields. It is this differentiation that appears to be the explanation of the U-shaped curve defining the ratio of size of holding to yield, which makes its appearance in a number of places.

An attempt to establish an over-all relationship between

yield per unit of land and size of holding under HYVs came to a somewhat indecisive conclusion. Of six studies in areas producing rice, three showed a positive and three a negative correlation; in two areas producing wheat, one showed a positive relationship and the other showed no significant correlation between the two; and in two areas producing maize, similarly there was one case in which yields were higher on larger farms than on smaller ones and one in which there was no significant correlation (*Dasgupta, 1977*). Thus, although in most areas a variable but large proportion of small cultivators continue to use local or perhaps nationally improved seeds, those who do move over to HYVs are likely to achieve reasonably high yields. The crucial factors are likely to be control of moisture on the agronomic side and ease of access to working capital on the economic side. However, their yield performance is in numerous cases matched and surpassed by large entrepreneurial cultivators.

What we have said above refers, of course, to cultivators with different sizes of holding who decide to adopt the new technology. But however attractive high yields may appear to the small cultivator, it is as an economist rather than as an agronomist that he will make his decision to change technologies, on the basis of his estimate of the net advantage he could expect to enjoy as a result of adoption.

The necessary calculations must take into account not only the level of anticipated yields but also the selling price of the harvest, the production costs incurred by the new technology, the costs of credit, and the possibilities of failure, leading to increased indebtedness or loss of land. It must also take into account additional managerial effort and annoyance and any constraints imposed by existing social obligations and prohibitions.

A large proportion of small cultivators, therefore, arrive at results from these calculations affording only a marginal, if any, net advantage and an insufficient incentive to take up the new technology. The less favourable the conditions, both agronomic and economic, for the practice of the new technology, the more generalized is its rejection by the small cultivator. This rule seems to apply as between regions, but also between cultivators within the same village.

TABLE 8

*Yields of Local, Locally Improved, and High-Yielding
Varieties* (kg. per hectare)

Place	Year	Local	Locally-improved	HYV
Rice				
Rai Rot, Thailand (well irrigated)	1971 (Wet)	5931	–	7908
Nong Serai, Thailand (moderately irrigated)	1971 (Wet)	3954	–	5931
Sa Krachon, Thailand (rainfed)	1971 (Wet)	2718	–	4448
Thanjavur, Tamil Nadu	1966–67	–	1903	3035 (ADT-2)
„ „	1967–68	2684	2805	3462 (ADT-2) Taiwan (3)
Ernakulam, Kerala	1966–67	1885	2545	3264
Krishna, Andhra	1966–67	3981	3951	4203 (TNI)
Kolaba, Maharastra	1966–67	1389	1446	1176 (TNI)
Cuttack, Orissa	1966–67	–	2100	4310 (TNI)
West Godavari, Andhra	1967–68	2829	3618	5689 (IR-8)
Birbhum, West Bengal	1967–68	2533	2857	3502 (IR-8)
Laguna, Philippines	1966	2493	2484	2329
„ „	1969	2849	3570	3841
Gujranwala, Pakistan	1968	–	2212	3687
„ „	1971	–	1567	2673
Wheat				
Ban, Karnal, Haryana	1968	1342	–	2281
Kota, Rajasthan	1968–69	7801	10836	10836
„ „	1971–72	12828	–	16352
Muzaffarnagar, U.P.	1966–67	1336	–	2414
„	1967–68	1962	–	3254
„	1968–69	2324	–	3482
Ferozepur, Punjab	1966–67	1873	–	2708
„	1967–68	1363	–	2950
„	1968–69	1562	–	2829
Maize				
Saran, Bihar		3294	–	7848

Source: Dasgupta, 1977.

And insofar as agronomic field successes of the new varieties of wheat have surpassed those of rice, so it is in certain
wheat-producing areas that, within the course of a few
years, 100 per cent adoption was achieved whilst the varied

performance of the new rice varieties, the spottiness of irrigation systems and the economic subjection of so many small rice producers have disinclined them to venture the technological leap.

II. PROFITABILITY AND SIZE

Non-adoption by small cultivators aware of the performance of the new technology indicates scepticism as to the benefits adoption will bring to their livelihood. In order to understand this, we must examine the profitability of the practice of the new technology. This involves a comparison of net returns per unit of land and per unit of output for the different size-groups.[2]

In eleven out of the fourteen sets of figures considered for India, costs per unit of output were lower for HYVs but in only four of those cases was the net return per unit of output higher for HYVs than for local and locally improved strains. In seven cases, the net return per unit of output was lower. However, net return per land unit was always greater for HYVs except where cultivators had to compete with the support price of the high quality basmati export rice (a locally improved variety) in Pakistan.

No systematic difference in market price of crops sold simultaneously was revealed as between small and large cultivators, but it was common for small cultivators to find themselves obliged in order to meet obligations (and in some cases for lack of storage facilities) to sell immediately after harvest (a period of market glut when prices reached their lowest), even where this meant buying back grain later in the year for consumption.

[2] Most of this data has been more closely analysed in Dasgupta (1977). In respect of market prices, the studies examined showed that on account of consumer preferences, the price offered for HYV grain was frequently lower than that given for local grain or locally improved varieties, to which consumers were accustomed. In the case of wheat, the price differential was not great and yield differential was higher than with rice, so that HYVs soon showed themselves to be more profitable whether by the unit of land or by the unit of output. However, the situation was less favourable for rice, the yield differential being less and the price differential often greater than with wheat. The result was that, although the net return per unit of land was greater, the net return per unit of output could be less than for locals.

The three studies of Northern India wheat-growing areas, where the success of the new technology has been most marked, show that following the introduction of the HYVs in 1967/8 there was a steady increase in expenditure on production costs per land unit from year to year amongst the large cultivators, reversing the initial situation, in which the smallest cultivators had highest production costs, due to the intensity of their use of inputs.

The economic performance on a per hectare basis of the large cultivator is seen rapidly to surpass that of the small in the four years during which the new technology spread through the area, though the situation of the small cultivator has also improved.

A similar trend is to be seen in the study of a newly irrigated village near Kota in Rajasthan (*Bapna, 1973*, p. 77), in which small (less than 15 acres) cultivator incomes increased by 9.3 per cent, those of medium (15–40 acres) cultivators by 11.9 per cent and those of large (40+ acres) ones by 50.1 per cent. These changes took place between the 1968/69 season and that of 1971/2.

In Bhalla's study (1972) of 723 farms from the whole area (1972) of the wheat-growing State of Haryana, the yield per acre continues to be in inverse ratio to size of farm, but with the larger adopters of HYVs already beginning to invest substantially from their savings in improved farm equipment.

The author (p. 35) takes the view that 'this trend in the long run is bound to lead to greater gains in productivity on bigger farms as compared to the smaller farms, and is also likely to result in greater concentration of capital stock with the larger farmers'. Cultivators of Category I (under five acres) he calls 'sub-marginal'. Only two-thirds of their income comes from their own cultivation, the rest being made up of wages from dairying and other sources. Their mean consumption amounts to Rs. 2,877, exceeding their mean incomes by Rs. 7. This level of consumption (at 1960–1 prices) corresponds to that recognized earlier by several authorities as 'the poverty line' (*Bhalla, 1972*, p. 17). More than a quarter of the cultivators in the State belong to this category and they appear to increase their debt annually. Category II (5–10 acres), accounting for a further 29 per cent of cultivators,

TABLE 9

*Gross Income and Farm Business Income for Various Farm
Size Groups, 1967/68, 1968/69, 1969/70, and 1971/72
in 3 Villages in the Indian Punjab*

Gross Income

(Rupees)

Year	Small	Medium	Large
1967/68	2320.75	1762.46	1911.93
1968/69	1779.82	1982.82	1096.01
1969/70	2208.69	2717.00	2452.29
1971/72	2574.79	2419.87	3925.40

Cost per acre

Year	Small	Medium	Large
1967/68	1481.09	1010.55	922.17
1968/69	1096.01	1233.82	991.82
1969/70	1230.55	1527.42	1285.87
1971/72	1439.33	1342.33	1637.39

Farm Business Income

Year	Small	Medium	Large
1967/68	839.66	751.91	989.76
1968/69	683.81	748.96	891.70
1969/70	978.14	1189.58	1166.42
1971/72	1135.17	1077.54	1658.01

Source: Dasgupta, 1977 (adapted).

obtain better incomes but still do not attain net savings, and
therefore cannot be expected to improve their productive
equipment significantly, 'nor are they in a position to withstand the vagaries of nature—drought, insect plagues or
floods'.

Only two of the studies available give data about input
costs by size for rice cultivation. One of these is a study of
commercial rice cultivation at Gurdaspur, Punjab (*Kahlon
and Singh, 1973*, I), and the other is the study done in
Province Wellesley, Malaysia, in which the variation in size
was minimal. The study done in Gurdaspur, like those in

wheat-producing areas of the same region, shows that inputs of fertilizers, machinery, and hired labour per unit of land increased with farm size, while bullock power declined. In contrast, the cultivators in the Malaysian locality studied are devoted mainly to self-provisioning in rice. Here, it is the cultivators of the smallest holdings (those of less than one acre) who spend more on fertilizers, insecticides and hired labour, and as much as the largest category on the powered tiller (*Bhati, 1976*).

Further evidence of the economic advantage enjoyed by larger cultivators emerges from analysis of the studies to determine optimum size of holding under the new technology from the point of view of efficiency; that is, the relationship of cost per unit of land to yield. Data from three wheat studies in Northern India were considered (*Kahlon and Singh, 1973*, II). In the case of Muzaffarnagar, holdings did not go beyond 26 acres in size and the cost curve was U-shaped, with costs per kilogram of the product working out at their lowest on farms between 11.7 and 17.2 acres.

In Kota, the size of farm ranged from 5 to 60 acres, and in this case two low-cost points were to be observed—one at 10-15 acres and the other between 40-60 acres, as noted above. These data point to the existence of two technologies, one based on bullocks and the other upon tractors. In Muzaffarnagar and in Kota, the optimum size for the bullock technology from the efficiency point of view was between 10 and 15 acres, and this corresponds with similar estimates made by Bhalla (1972) and Khusro (1973). However, for those who had mechanized, there was an optimum efficiency level at 40-60 acres.

In Ferozepur (*Kahlon and Singh, 1973*, II) the cost curve showed continual decline for an area where tractors were rapidly replacing bullocks. It is assumed that the easy credit facilities offered to aid mechanization had led many cultivators to adopt the use of tractors prematurely.

The only rice studies that provided suitable data for analysis of efficiency and profitability in relation to size were three done in Sri Lanka and in each of them it was farms of the 10-15 acres category that provided the highest level of net return per unit of output, while in two out of the three cases, farms of this size turned out to be the most efficient in the sense of having the best cost/output ratio. Equally, in two of the studies this category attained the maximum net return per acre, and in the third it only just fell short of the maximum.

Concentration

In discussing the size of farm in irrigated wheat production, we are obliged to draw the conclusion that in itself, small-ness of farm was no serious obstacle to high yields under the new technology. In the case of North-West Mexico, the Green Revolution took place on land carved out of the deserts by government-financed irrigation schemes, and technique was mechanized from the start. However, as serious farming spread, a surprising process of concentration took place.

In 1948, there were rather more than 400 'large' farms in the Hermosillo coastal region with irrigated lots of 100 hectares apiece. By 1956, with the rapid extension of existing farms into the desert, the irrigated area had increased by 75 per cent, though the number of farms was only 260, with an average of 267 hectares each. In 1971, a national finance commission attempted to find the 1,000 titular owners of private farms, and after extensive researches concluded that there were only 150 enterprises extant at that time, holding approximately 800 hectares each on the average, though in fact sizes were dispersed. Similarly, private farming in the Yaqui Valley was said to be concentrated in the hands of 106 owners averaging perhaps 500 hectares each. Indeed, the concentration had gone so far that 'the largest landowners of Hermosillo are also the largest landowners of the Yaqui Valley and frequently of Caborea, the Altar region, and the principal irrigation districts of Sinaloa' (*Hewitt de Alcántara, 1976*).

The vigorous expansion of individual farming enterprises using the new technology with a high level of mechanization did not stop at the purchase of land. Hewitt reports that in a 1963 survey one-seventh of all lands operated in the Yaqui Valley was found to be rented, the big farmers leasing in lands from those who were unable to use the new technology profitably, i.e. smallholding *colonos*, members of communal landholding groups (*ejidos*), and members of Yaqui Indian communities (*Hewitt de Alcántara, 1976*).

Readers of the report on the Mexico country study will see that development of modernized agricultural production in North-West Mexico had a sharply dichotomizing effect on existing farming communities and brought about a deep separation between those who lived from the sale of farm labour and those who lived from the control of capital. In crowded Asia, the existing widespread diffusion of land-owning provides resistance to excessive concentration of ownership. Nevertheless, incipient trends are to be observed in some of our studies in the new technology areas.

Apart from North-West Mexico, the optimum area for the new technology is to be found in the level, irrigated lands of the Sub-Himalayan plains in North-West India and Pakistan. The Indian sector corresponds to the Punjab, Haryana, and the western side of Uttar Pradesh. Global Two studies were carried out in wheat-growing districts of these three States (respectively in Ferozepur (*Kahlon and Singh, 1973,* II); Karnal (*Laxminarayan, 1973*); and Muzaffarnagar (*Singh, 1973*)). A further study was done in a region of expanding canal irrigation where very similar conditions prevail, namely at Kota in the neighbouring State of Rajasthan (*Bapna, 1973*). An additional survey covering farms in Haryana (*Bhalla, 1972*) was also available.

In Ferozepur, between 1967/8 and 1971/2, the average value of land in the sample studied rose from Rs. 138,000 to Rs. 242,000 (a rise of 75 per cent) and, in spite of increased investment in machinery, land still accounted for 84 per cent of the total value of assets.

In the four Haryana villages studied, the three in which the new technology was successful show increases in the price per acre of irrigated land between 1966/7 and 1971/2 of approximately Rs. 6,000 to Rs. 10,000, Rs. 4,000 to Rs. 10,000 and Rs. 5,000 to Rs. 8,000 respectively, while the village in which the new technology had not caught on showed an increase of only Rs. 4,000 to Rs. 4,500. (*Laxminarayan, 1973*.)

In Kota, where large-scale irrigation was in progress and the irrigated land area in process of expansion, the price of land rose 20 per cent. (The factor that seemed to be restraining the price of land in Kota was the fact that supplies of irrigated land were still becoming available. Indeed, much of the land purchasing brought to light by the study was done by cultivators from the areas of rising land prices, especially from the Punjab.)

In the Muzaffarnagar study, the rise in the value of land between 1967 and 1972 was of the order of 185 per cent, going up from Rs. 5,262 to Rs. 15,012 per hectare.

The figures for Ferozepur and Karnal also show that the increased demand and higher price of land favoured its concentration in the hands of the already large landowners. Small proprietors, who constituted the majority of cultivators, could no longer invest in the purchase and improvement of land while the larger farms invested heavily in land, once the possibilities of the new technology became apparent.

The increased land concentration in the Haryana villages is shown clearly in Laxminarayan (1973). In the three villages where the new technology was adopted, it is the largest proprietors whose gains are greatest, while in the village unaffected by the new technology (Mirka) the trend is not followed.

In Ferozepur, the average size of holding for the three size groups changed in a similar direction, the small proprietors' average size moving from 4.80 to 3.47 hectares, the medium from 8.96 to 7.25 hectares and the large from 15.63 to 18.08 hectares.

Finally, figures for the villages in Kota district, Rajasthan, show that the only size group to make any substantial advance in 'area operated' was in the 'more than 24 hectares' category, whose average rose from 92 to 106 hectares.

The same trend, involving the ousting of marginal cultivators and the establishment of entrepreneurial cultivators in their place, is to be seen in the case study showing changes in Muzaffarnagar, Uttar Pradesh.

Highly Profitable

The most striking fact demonstrated by these varied data is the appearance of the mechanized commercial wheat producer in the irrigated lands of North-West India and the neighbouring areas of Pakistan. It has taken but a few years to learn that with adequate investment in the prescribed seeds, fertilizers, and protective chemicals, with level land and tubewell irrigation providing the essential foundation of crop security, and tractors to keep labour costs and uncertainties to a minimum, a wheat farm of 16–24 hectares can be a highly profitable enterprise. It is also demonstrated that in this area, where moderately reliable irrigation is available, even very small farms of less than 2 hectares which, under pre-HYV conditions needed supplementary incomes for providing family livelihood, are able to achieve a high level of yield, provided credit is available to them for the high operating costs, for seeds, fertilizer, and chemicals.

However, there are certain contrasts between the small and the large cultivators as defined here that are likely to be critical in determining future trends. Beyond a certain level in landowning (approximately 4–6 hectares, at which size-level the economies of scale for bullock plough users are likely to ensure the highest net gain per hectare), the cultivator will begin to accumulate from each crop net gains that are not absorbed by his domestic requirements. These savings can be applied directly to his own operations in such a way as to further increase net gain. By increased self-financing, he can avoid heavy interest payments.

Savings also render distress sales of grain immediately after harvest unnecessary and some investment in storage or renting of storage space can make possible the disposal of the market surplus when conditions are profitable. The purchase of machinery such as sprays and tractors is also cost-saving. With increasing profits, land may be rented in or even purchased strategically so as to obtain maximum benefits from the lumpier (less divisible) investments in machinery and installations. In fact, the cultivator may enter a path of dynamic growth.

It is very unlikely that this path of growth will be open to the very small cultivator: the expansive successes of his larger neighbour may even block his entrepreneurship. The increased price of land makes it difficult for him to expand his holding either by purchase or by renting, while it makes abandonment of farming and the sale of his land more attractive. It looks as if the growth of entrepreneurial farming will tend to edge out small operators in spite of the scale-neutrality of the genetic-chemical components of the input package.

In the long run, the squeeze is likely to come when price reduction follows the cost reductions possible for the mechanized entrepreneurial farm, leaving the small operator unable to compete. This is not to suggest that anything so drastic as the collapse of small-scale farming in the Yaqui Valley is imminent in the Punjab. But it is realistic to examine what kind of political power is exercised by the new entrepreneurial farming class and its commercial allies, and the agricultural development policies a State Government based on the support of this sector might be expected to pursue.

It was possible to see in relatively clear outlines certain of the processes set off by the large-scale introduction of high-yielding varieties of wheat in North-West Mexico and the Sub-Himalayan Plain in both India and Pakistan because of the drastic changes that took place in these regions as cultivators began to use the new seeds and to establish the conditions in which their high-yield potential could flourish. Although some of the new varieties of rice have had remarkable agronomic successes, the picture of the changes following the large-scale introduction of the new technology in rice is much less decisive, more uncertain, varied, and localized.

Rice is a small cultivator's crop in most Asian countries, with 49 per cent of all operating units in South-East Asia being of less than one hectare and a further 36 per cent being of 1–4 hectares. It is perhaps significant that the most successful application of the new technology in rice-growing seems to have taken place in rather a small area of the Punjab, under socio-economic and geographical conditions similar to those in which wheat is produced. Sri Lanka has also developed yields steadily, achieving a 52 per cent increase between 1952–4 and 1967–9 by promoting a technology based on varieties produced by its own research centres.

Unquestionably, for the small cultivator perhaps the main obstacle to the effective deployment of the new technology in Asia has been its high production cost of obtaining credit. The producer with one or two hectares of paddy has to face especially great difficulties because a portion of all of the land he cultivates belongs to someone else. To understand his

position, therefore, we have to examine the economic difficulties of the would-be entrepreneur with neither land nor
capital of his own.

III. IS SMALLNESS AN ECONOMIC HANDICAP?

It is commonly said that cultivators holding less than a
certain minimum of land are liable to diseconomies of scale
and are more vulnerable to risk than cultivators with larger
holdings. Although the main features of the new technology
are scale-neutral, in the sense that seeds and chemicals can be
procured in small quantities suitable for small land areas, the
small cultivator suffers from an accumulation of minor diseconomies regarding other inputs.

The most obvious of these is draught power for tilling and
transport. A pair of oxen, for example, may be an essential
part of the cultivator's capital investment, yet the animals
must be maintained the whole year round for only a limited
number of work-days. He may also have a similar problem
with tractors, transport vehicles and other machines, so that
unit costs are greater.

A further problem for the small owner-cultivator relates to
water. Should he receive canal water or operate his own well
with animal power, he may enjoy arrangements suitable to
the size of his land. If, however, he must rely on a power-
driven pumping system that is not publicly owned, he is
likely to be forced into dependence on a monopoly supplier.
In this case he would be obliged to meet heavier production
costs per land unit than the larger landowner, who can invest
in a powered unit fitting the size of his holding.

Management tasks connected with the new technology
such as attending meetings, presenting credentials for obtaining loans, and going to the nearest town to ensure receipt of
inputs or to arrange for marketing tend to be disproportionately time- and money-consuming for the small cultivator.
His livelihood is based on his own labour, whether in his own
fields or in return for a wage in the fields of others. None of
these additional difficulties should be exaggerated, however,
since there are ways of overcoming them so as to minimize
the disadvantages.

It will not escape notice that while the genetic-chemical technology is in itself almost scale-neutral, its very success is likely to push the larger cultivator into mechanized techniques that are not.

The new technology also introduces causes of risk additional to those of crop failure due to natural causes, i.e. weather, disease or pests. The price paid for the product may fall below expectation and the price paid for the inputs may rise above what was expected, so the innovator could lose or gain more than did the producer of local varieties (*Griffin, 1972*, pp. 58–9).

Griffin doubts that the small farmer is more averse to risk than the large one. What must be recognized, however, is that the small farmer or tenant may take more drastic risks, putting at stake his whole livelihood. The cultivator with a larger farm property risks commercial losses but his reserves enable the enterprise and the livelihood it supports to survive a few years' crop losses. These economic disadvantages are certainly not overwhelming, and they can be remedied (provided the political will is there) by a variety of institutional measures and services. It is sufficient to quote the agronomic and over-all economic success of small cultivators in Japan and Taiwan to realize that the small farm cannot be isolated and condemned as uneconomic.

The problem is rather that of economic polarization in market economies where small-scale peasant agriculture must compete with a rapidly growing capitalist farming sector. Moreover, it cannot be adequately explained in terms of the economies of size.

VII CHANGES IN ASIAN TENANCY

*The importance of tenancy is demonstrated—both as the
system that affects the lives of most Asian peasants and
also as one of the means by which they are exploited.*

Tenancy involving patron–client structures has been characteristic of the recent agrarian history of most Asian countries. It
is a feature that was accentuated, used, and sometimes
imposed by the colonial powers as a functional system for
the collection of taxes from the primary producer. The
tenant cultivator was dependent on the dominant landowning
families and his labour maintained them in their privileged
situation, but, at the same time, the capacity of the richer
families to accumulate wealth and to maintain certain facilities such as storage, processing plant, or water tanks, provided
some security for the tenant, who was indispensable to his
landlord's welfare.

The advent of the new technology is part of a larger
change process that has been gathering force rapidly in the
last decades: the penetration of the village by market forces
and its incorporation in the hurly-burly of larger competitive
market systems, as unpredictable in their demands as in their
offerings. As the Green Revolution brings higher levels of
mercantility and as the profitability of the new technology
for the well-funded entrepreneur is clearly demonstrated, the
raison d'être of the traditional tenancy arrangements is
undermined, and the patron–client relationship becomes a
tiresome encumbrance to the landowner intent on a more
dynamic use of his capital.

Meanwhile, the bargaining position of the poor cultivator,
which at an earlier stage was real enough, sinks as a result of
the relative as well as the absolute increase in the number of
landless and land-poor families in most villages, hungrily
competing for work, for wages, for grain or for land whatever
the conditions.

So with higher land prices, landowners are to be seen
choosing between three main options. For some the moment

is an appropriate one for selling their land at a favourable price. Others commit themselves more fully to entrepreneurship and the new technology, becoming cultivators and using the security of their property to obtain credit for machines and for the new higher costs of production. Finally, there are those who are not in a position to take up personal management of a farm, but who wish to get as large a share of the profits to be anticipated from the technology as possible. How the third of these options is taken up is illustrated from the countries studied, and especially from India.

I. TENANCY IN INDIA

During the colonial period in India, there were various types of land tenure, with tribal arrangements in the remoter and less populous parts. The *zamindar* system maintained a complicated pattern of tenancy serving the colonial purpose of tax-farming. In addition, small owner-cultivatorship was encouraged and showed a certain amount of vigour, though there was little technological development in spite of the rapid changes taking place in Europe.

Khusro (1973, p. 5) attributes the technological stagnation of Indian agriculture, especially in the last decades of the colonial period, to the fact that the cultivator himself was never able to accumulate savings and invest in the improvement of his land and tools, and to the severity of the demands made by the colonial land revenue system (in smallholding— *ryatwari*—areas), as well as the arbitrary exactions of landlords and intermediaries under the *zamindar* system.

Following Independence, the *zamindar* system as such was abolished and a large-scale transference of property in land took place, both to actual cultivators[1] and also to investors who sought a secure rental income. Whilst the renting out of land was regularized and submitted to legal procedures and measures to protect the tenant, these measures also served to drive tenancy underground.

---

[1] The phrase 'actual cultivators' does not necessarily mean 'working cultivators' in India; outside the *ryatwari* areas it is common for even small cultivators to have their field work done by labourers working by the day or annually contracted.

An anthropologist (*Mencher, 1974*) reports that one of her student assistants 'spot-checking in 1972, in the villages where she had worked in 1971–2, noted that *none of the tenants* had been given any permanent right to cultivate . . . plots of land as provided in the legislation'. The proprietors defended themselves against possible legal action by changing their tenants regularly so that they could not claim permanent rights as a result of several years' occupancy.

Many non-working landowners dubbed themselves 'owner-cultivators' and designated their tenants as 'servants'. By thus avoiding registration of the tenancy, landowners were also able to avoid the protective provisions for ensuring a fair rent. 'Owing to this motivation', says Khusro (1973) '. . . crop-sharing arrangements are more often than not shrouded in a veil of secrecy. Absentee landowners either have un-recorded tenants or else they make a servant's deed with their tenants rather than a lease deed, although on the application of the acid test the "servants" appear to contribute a sub-stantial portion of farm capital if not the whole of it.'

Nevertheless, the over-all tendency in the period prior to 1966 seems to have been one of marked decline in the area under tenancy in most regions, and a more limited decline in the number of tenants, with new commercial tenancies appearing at a higher level. Comparing the results of the Eighth and Seventeenth Rounds of the National Sample Surveys, Narain and Joshi (1969) claim that between 1954–5 and 1966, the area under tenancy fell from 20.34 per cent to 10.7 per cent, the decline being relatively more pronounced amongst the large holdings than the small ones. Support for this trend is found in community re-studies, 10 of which are quoted, from Uttar Pradesh, West Bengal, Punjab, Orissa, Andhra Pradesh, and Tamil Nadu. In only one case is it mentioned that former tenants acquired possession of the land. In most of those mentioned, the significant change is the withdrawal of larger plots of land for personal cultiva-tion by the owner, in response to improved prices for cereals and *gur* or coarse sugar. (*Narain and Joshi, 1969.*)

However, the significance must not be misunderstood. According to some estimates, the majority of cultivators (the poor) till only 13 per cent of the total area in farmed land, so the NSS figure of 10.7 of the area could still hold a large number of tenants. As we shall see from the examples, tenancy is very widespread still, and the reality of its magnitude certainly does not appear in the published statistics and may have eluded the community studies and re-studies as well. As it is also true that large numbers of tenants have lost their lands to entrepreneurial landowners, there is no reason to believe that new share tenancies or share-cropping arrangements are not constantly being entered into.

Many state land reforms established a property ceiling but made generous allowances to those landlords who took back their property for 'self-cultivation', and these measures also encouraged the recovery of lands from tenants, who could in some cases be employed as workers, but on conditions very little different from those of tenants.

Thus the new technology began to make itself felt at a time when the institution of tenancy was in a state of flux. As a result of improved prices, and the regulation of tenancies by law, owners were withdrawing middle-sized and larger parcels of land from tenants. They were difficult to conceal, and the larger tenants could not so easily be made a party to disadvantageous subterfuge. Also, the level of rent that could be charged did not compare favourably with the anticipated gains to be obtained from full entrepreneurship and direct investment. This tendency, as we shall see, was replaced by the vigorous expansion of larger and well-capitalized operational holdings by recourse to commercial tenancies at fixed cash rents in those optimum areas where the high-yielding varieties demonstrated a high level of profitability.

II. INDIAN CASE MATERIAL

In view of the inconclusiveness of statistics, let us now examine field material from studies done by different research groups whose members took the trouble to find out what was happening to tenants and tenancies in villages and areas in six States where the new technology was being introduced on a large scale.

(i) Ludhiana, Punjab: landlords raise rents and tenants are unable to meet production costs of new technology

The position of tenants in the district of Ludhiana, Punjab, which may be considered as the heartland of the Green Revolution, is described in some detail by Francine Frankel (1970) as the potential profits from direct cultivation look increasingly attractive, and the demand for land rises. Where large proprietors still give out land on lease, rents charged are 50–70 per cent higher than before the introduction of the new varieties, but share-tenancies are more common.

In many instances, landowners may ask for 70 per cent of the crop as their share, arguing that with the new methods the tenant still receives a larger absolute portion from 30 per cent of a higher output than 50 per cent of a lower out-turn. But since the small owner-cum-tenant cultivators cannot afford to invest in optimum cultivation practices, they find the new rentals uneconomic and generally are forced to give up . . .

Another alternative mentioned by Frankel can best be described as 'share-labourer', in which a large owner provides land and inputs to a cultivator with or without resources, to whom he allows 20 per cent of the harvest in return for his cultivatorship and labour.

(ii) Pedapulleru, Andhra: gains from new technology appropriated by landlords

In the rice-producing village of Pedapulleru, Andhra, 10 per cent of the population owns 80 per cent of the land, and of the 42 per cent of cultivators' households approximately half are tenants. The control exercised by those who have a quasi-monopoly of land is strengthened by the fact of their being members of an hereditary landowning caste whose internal solidarity and endogamy prevents the wider diffusion of property holding common in peasant societies. Under such conditions of resource control, it is possible for the large landowners to adjust the terms of tenancies so that the benefit in yield and profit accrues to them rather than to the tenant. This is described as follows by Parthasarathy (1973):

Two types of rental payments are in practice in the village, viz., share rents, and rents fixed in kind. Rental systems vary between the Kharif and the Rabi seasons. For traditional varieties the most common practice in Kharif is payment of a fixed contracted amount of produce in kind, and in Rabi a contracted share in the gross produce. When the rent is fixed in kind, all the cultivation expenses except land revenue are borne by the tenant. Then it is paid as a share of gross produce, the costs of fertilizer and pesticides only are borne by the landowner proportionate to his share in the gross produce. Since 1965/66 there is a slight swing towards share tenancy, even in respect of traditional varieties in Kharif. In Rabi it was always a share of produce, and it continues to be so. The major change is now in the level of share paid to the landowner when HYVs are grown.

When traditional varieties are grown a share of two thirds of the produce is given to the landowner only in a minority of cases. But if the tenant chooses to grow HYV the landlord insists upon the two-thirds share and 10 out of 15 Rabi HYV tenants pay two thirds of the gross produce. Of course, the landlord shares two thirds of the cost of fertilizer and pesticides. Thus as the production potential increases, rents are revised upwards, and the rental market does not automatically ensure the flow of benefits of technology to the tenant. Decision-making also shifts more and more to the landlord. He decides the variety to be grown, supplies a major part of the capital for non-traditional inputs and provides the finances to the tenant. The tenant becomes indistinguishable from a permanent farm servant and the tenancy system nothing but a convenient arrangement under which a big owner relieves himself of the burdens of labour management while performing the major entrepreneurial functions. The new role played by the landowner is in a very large measure attributable to changes in technology, and this provides an explanation for the observed similarity in the level of inputs as well as yields between owner and tenant groups.

Rents paid per hectare are studied for each season and within each season, for traditional rice and HYV separately. Rents paid for area under HYV are more than double the rents paid under traditional crops in the Rabi season. Even in Kharif rents paid for HYV are more. The quantum of rents paid under share rents is found to exceed that fixed in kind. If allowance is made for the sharing of expenses between the landowner and the tenant, the gain of the landlord in the total produce under the two arrangements is more or less similar. Considering the Rabi season and pure tenants only, the yield is 27.21 quintals per hectare for local seeds as against a rent of 14.44 quintals. The yield under HYV is 57.27 quintals as against a rent of 37.05 quintals. By a shift to HYV the tenant gets an excess of 30.06 quintals per hectare and pays 23.41 quintals more by way of rent, thus gaining an excess of 6.65 quintals. Even after making allowances for excess cash costs, the tenant is a gainer. But if relative shares are examined, more than three quarters of the excess produce is got by the landowner and a quarter by the tenant, making the relative distribution of income much more uneven.

(iii) Purnea, Bihar: Unlawful tenancy conditions depress both livelihood and productivity potential of new technology

A close-up of one of the forms taken by share-tenancy is given by Bell and Prasad (1972), who selected a sample consisting of 25 pure share-tenants and 31 owner/share-tenants from villages in the district of Purnea. Considerable importance was given to the question of the legal recording of tenancies, since they were subject to legislation both as regards security of tenure and as regards contractual conditions. Of the sample, 30 of the 56 arrangements were completely unrecorded, and in a further 13 cases, recording had been carried out in relation to only a part of the land held in tenancy. In only 13 cases were tenancies fully recorded. However, even in these cases, the legal condition that the product should be divided in the proportion 75:25 was not stipulated.

So long as a locality-bound production technique was used, the only inputs were seed and perhaps irrigation and compost-manure, and these, though not large, had to come out of the tenants' 50 per cent, along with the cost of any labour that might be hired additionally to that of the harvesters. The tenant also had to supply the instruments of tillage, bullocks, and other fixed capital.

Table 10 shows that landlords receive about 46 per cent of the value added with both types of tenant, that the owner/share-tenants use more hired labour than those who have no land of their own, and that the pure share-tenants use more family labour and less hired labour than the owner/share-tenants, who use the same amount of hired labour on their rented-in lands as on their own.

In addition to the production of cereals, the livelihood of these tenant-cultivators is supported by some animal husbandry and agricultural wage labour, plus a small amount of trade and some transporting, especially on the part of the owner/share-tenants. Their average annual family income expressed in per capita terms worked out as Rs. 210 for the latter and Rs. 145 for the tenant families without land of their own. On a household basis the figures were Rs 2,025 and Rs. 1,225. Had the law been observed, and the product

divided 75:25, then the authors calculate that the per capita
income figures would have reached Rs. 245 and Rs. 195
respectively, on the conservative assumption that reductions
in the rental share would have no effect on farming intensity.

TABLE 10

*The Distribution of Value Added from Crop Cultivation
by Share-Tenants in the District of Purnea, Bihar*
(in rupees)

	Landlord	%	Hired Labour	%	Tenant	%	Total
Share-tenants	25,092		8,462		21,969		55,523
		45.1		15.3		39.6	
Owner/share-tenants							
a) on rented-in land	23.213		11,673		14,627		49.513
		46.8		23.6		29.6	
b) on own land	—		8,333		31,322		39,655
		—		21.0		79.0	

Note: The figure in the lower right-hand corner of each cell shows the percentage
    of total value added received by each class of income receiver by tenant
    category and tenure status of land operated.
Source: Bell and Prasad, 1972.

If the tenant were to become owner of the land, paying it
off in about twelve years at the usual land mortgate rate, for
those years he would be paying about Rs. 185 per year, not
very different from the value of the usual 46 per cent landlord's share.

Both output and income are also likely to improve as a
result of improved allocation of resources.

The existing 50:50 division is a serious obstacle to use of
the new inputs since the full cost of these, as well as the
burden of risks of crop failure, must fall on the cultivator. He
is, therefore, likely to limit himself to reduced applications of
fertilizers and other purchased inputs. If the product were
split 75:25, then a fuller use of these inputs could be expected.
It follows that such a readjustment would lead not only to
greater equity but also to improved output and higher
incomes for the cultivator.

A further suggestion is that when the new inputs are used, a cost-sharing/harvest-sharing arrangement should be introduced, with the landlord contributing to costs of production in the same proportion as he shared in the harvest.

According to Bell and Prasad, some cases have been observed in which landlords share 50 per cent of the cost of one or more input (for instance, irrigation) but these cases were mainly found among the commercial medium-size share-tenancies. New legislation enforcing cost-sharing where new inputs are concerned, say the authors, might be expected to provide a powerful impetus to an existing if still feeble trend.

They go on to propose that on the basis of a 75:25 cost- and harvest-sharing scheme, the institutional credit allowed to the registered share-tenant should be repaid 25 per cent by the landlord.

A fixed cash rent is also proposed as an alternative. In this way it would become a fixed cost of production, so that decisions concerning input use at the margin, for profit-maximizing behaviour at least, are unaffected by its presence.

The difference between the two modes (cost-sharing/harvest-sharing tenancies and fixed rent tenancies) resides in the risk-sharing nature of the first, so the riskiness of the particular crop might be a decisive factor.

Both forms are desirable because by making for greater efficiency of resource use and hence for rising output, 'they give more room to manoeuvre. The tenants' present consumption level can be improved alongside his ability to purchase the land outright' (*Bell and Prasad, 1972*).

The main obstacle to improving the system is the excessive difficulty in getting share-tenancies 'recorded'. Land records are so deficient and the measures needed to bring them up-to-date so cumbersome, dilatory, and expensive that they take years to materialize. The authors propose starting with special *ad hoc* campaigns in selected districts.

(iv) Thanjavur, Tamil Nadu: Excessive fixed rents and high interest limit tenants' operating capital

The district of Thanjavur in the State of Tamil Nadu is an area of rice monoculture covering 1.5 million acres, of which 86 per cent are irrigated. Its population density reaches 335 persons per square kilometre. According to the 1961 Census, which was based on a 20 per cent sample, 42.5 per cent of the total area was operated by owner-cultivators, 29 per cent by tenants and 28.5 per cent by owner-cum-tenants. Abdul Hameed[2] took a random sample of 420 cultivators from all the 36 development divisions of the district and he arrived at an estimate of slightly less than half of the area farmed as being rented-in land. Thus, the comportment of these cultivators is very significant for over-all production in this rice-bowl and, at the same time, the level and quality of livelihood they are able to achieve must be considered as representative of the middle levels of the rural population in that area.

Ninety-five per cent of the tenancies are according to *kutthagai* arrangements, which oblige the tenant to pay a fixed rent in kind after the harvesting of his crop. The landowner pays the land tax but does not participate in production costs. Minor improvements in the tenants' cultivation practices had not led landowners to raise rents, which were already very high, being between 51 and 60 per cent of the gross value of the product. The reasons given for adherence to fixed rent arrangements were several. More than one-fifth of the owners of leased lands were trusts, especially religious bodies, and of the remaining individual owners, 14 per cent lived abroad, 43 per cent in towns and cities, 36 per cent in distant rural areas, and only 7 per cent of the owners lived in or near the village where the rented land was situated.

The payment of fixed rents was therefore more convenient for the great majority of landowners since these were absentees and corporate bodies, and consequently not well placed for supervising the division of the harvest. In addition, it was more convenient for the trusts to be able to look forward to realizing a fixed sum each year rather than receiving a variable share. From the tenants' point of view, the special characteristic of the share rent—that it diminishes in case of crop failure—was not attractive since under the existing system of canal irrigation crop failures are infrequent.

It might be reasoned that because the tenant cultivators

---

[2] This account of tenancy in Thanjavur is based on *Hameed, 1971.*

could look forward to enjoying the full benefits of an increase in output due to the use of bought inputs such as improved seeds and fertilizer, they would have proved to be ready adopters of the new technology. But such was not the case, and both input use and yields of tenants were significantly lower than those of owner-cultivators. Data given in Table 10 show that the family labour input of tenants was consistently higher and the amounts spent on bought inputs were consistently lower for tenants than for owners.

Tenants have also shown themselves loath to intensify land use by undertaking a second crop, as is manifest by their lower rate of cropping intensity—129 per cent as against 141 per cent achieved by owner-cultivators.

How are we to explain their poor performance?

Low input use must be put down to the tenant's difficulties in obtaining working capital at a rate that would not endanger his anticipated net gain. Hameed ascribes this to (i) the high rents themselves, which prevented the tenant from making savings that could be retained as working capital and (ii) the practice of the credit co-operatives, to which most cultivators belonged.

These co-operatives were less discriminatory than most of those in India on which we have data: they enrolled members irrespective of tenurial status and gave production credit without collateral, even to tenants who cultivated land on oral leases. But the tenant is allowed only 60 per cent of the amount of credit per acre advanced to owners, and he is thus obliged to enter the informal credit market for his remaining needs. And the records show, as could be expected, that the tenants obtain a higher proportion of their credit from moneylenders, and pay higher rates of interest for it, than the owners do.

Capital is scarcer and costlier for a tenant than for an owner-cultivator, so tenants restrict its use to a point short of that which equates marginal cost and marginal value product of capital in owner-cultivated farms. Thus the dearth of capital resulting from credit rationing practised by co-operatives and private lenders would seem to explain in part the low capital intensity observed on tenant farms.

High rents and high interest withdraw from the tenant the

capacity to use net gains as a contribution to the production costs of the next crop. And the reason given for the tenants' lack of interest in multiple cropping seemed to be that wage work was available at the time when the second crop would need his own or his family's labour and the net benefit to family livelihood from wages earned outside was substantially higher than that which might be anticipated from a second crop tenancy contract, for which fixed rent of approximately 47 per cent of the gross product would be charged.

In effect, extortionate pressures are exerted both by the owner of the land and by the moneylender that depress the remuneration of the tenant for the labour of his family and his animals to a tolerance level determined by the prevailing economic condition of the region and the livelihood alternatives available. This phenomenon may be called the 'surcharge for survival', and it is one of the mechanisms that determines the persistence of poverty in a market society where neither community solidarity nor organized social welfare soften the edge of competition.

The force of demand becomes infinite when the goods demanded are indispensable. 'Stand and deliver! Your money or your life!' says the highwayman, and the unarmed passenger hands over his whole surplus yet retains his life. Those who deal with the poor over the ultimate elements of livelihood know that they can exact this ransom.

(v) Villages in Gujarat: Increased demand for tenancies tilts conditions against share-tenants

The case for the shrinking of tenancy rests mainly on evidence of the decreasing supply of land for tenancies. The argument is that as the profitability of agriculture rises as a result of improved technology, landowners are inclined to become cultivators, and therefore to withdraw land from the 'tenancy market'.

The persistence of tenancy is to be found where both supply and demand for tenancies are sustained, and it is interesting to look at the cases in which this happens.

Vyas (1970) provides an illuminating example from the study of a number of villages in Gujarat of a situation in which the introduction of the new technology, and other factors raising the capital-intensity of agriculture and augmenting the return on cash invested, have a dynamic effect on tenancy. The new trends can be analyzed in terms of changes taking place in both supply and demand factors in the 'usufruct market'. Two special factors in the situation studied here are:

—the legally imposed ceiling on land ownership, which, though in many respects ineffective, does inhibit large landowners from further purchases of land;

—the proximity of a thriving urban centre, which acts as a magnet for migration from the village by owners of land both large and small, and feeds distinct tenancy markets.

On the one hand the large landowners prefer to lease out to relatives who are medium or large cultivators, and who become commercial tenants paying a fixed cash rent. Vyas also mentioned the acquisition of tenancies by large cultivators from small by means of 'usufructuary mortgages', that is, cash loans made against the usufruct of a parcel of land belonging to the borrower, usually a small landowner. In addition to the market for commercial tenancies, there is a second market in which owners of very small parcels of land seem to maintain a precarious economic balance by supplementing their own land with small plots of leased-in land. This market is greatly affected by the increasing demand for land by commercial tenants, with the result that supplementary tenancies required to reach the livelihood threshold are harder to obtain, and cost more in rent.

Vyas was able to get information from eleven share tenants in two of the villages for the 1967–8 *kharif* season, and found that the four landless were paying an average rent imputed at Rs. 162 per acre; five small landowners with less than 2.5 acres were paying an average of Rs. 141, while two landowners in the 2.5–7.5 acres category were paying only Rs. 98.

Comparing the two traditional technology (TT) villages with the new technology (NT) villages, he shows that the size of operational holdings had become much more skewed in NT villages, where 58 per cent of the owned land and 44.4 per cent of the leased-in land was worked by the top 20 per

cent of landowners, while in the TT villages the corresponding figures were 38 per cent owned land and 16.7 per cent
leased-in land worked by the comparable 20 per cent.

The pressure upon the small cultivators was even more
strikingly illustrated by the fact that the smallest 20 per cent
in the two NT villages had 87.3 per cent and 73.9 per cent
respectively of their total operational area composed of
leased-in area.

Thus, the tendency to owner-cultivatorship, the new
profitability, and the existence of land-ceilings all contributed to an increased demand for commercial tenancies,
while supply was fostered by the alternative job opportunities
for small landowners in the growing urban areas nearby. It
was not possible to tell from the data given what net employment situation resulted.

III. TENANCY UNDER STRESS IN SRI LANKA, INDONESIA, AND MALAYSIA

(i) Sri Lanka

We have already noted the decline of owner-cultivatorship of
rice in the Philippines, and the prevalence, even more intense
in the optimum rice areas of central Luzon, of a special brand
of petty tenancy, *kasamahan*.

A similar trend is to be seen in Sri Lanka, where there was
a decline in the proportion of owner-operator holdings from
61 to 47 per cent between the Census of Ceylon, 1946, and
Census of Agriculture, Ceylon, done in 1962. And in only
one of the four locality studies, that of a new land settlement
scheme, was tenancy negligible.

In Minipe settlement, although the emergence of a populous new generation of cultivators had fragmented holdings
and decreased their size, and though the new technology had
fostered the emergence of an economic élite, farming had
reached a high average level of productivity. It brought with
it standards of livelihood that compared well with other
(including urban) occupations (*Hameed, 1977*). In Ussapitiya,
it was found that *ande* (share) tenants were not registered as
such under the Paddy Lands Act, but their landlords were,

though the latter were not working cultivators. The Act was, therefore, inoperative and the tenants paid 50 per cent of the crop in rent instead of the legally stipulated 25 per cent. In fact, one tenant had taken legal action against his landlord and had his rent reduced to 25 per cent of the crop. Other tenants declined to take legal action, they said, 'for fear of hurting the feelings of their landlords'. In view of the high level of landlessness and unemployment in this village, such delicacy was prudent.

The *ande* system of share tenancy as practised in Ussapitiya stipulated the division of the crop into two equal halves after the repayment of the seed grain (at 1.75 for every 1 measure), with all other production costs to be paid by the tenant. No institutional credit was available in this village, although cultivators could have asked for it had they known that they were entitled to receive it. Of 102 cultivators in Ussapitiya, 53 were *ande* tenants, 41 were owner-cultivators and 8 were tenants on fixed rent. Ten of the owner-operators were from outside the village, as were 27 *ande* tenants. Presumably most of these worked land belonging to outsiders, since two-thirds of the paddy lands had been sold by their original owners to persons living outside the village.

In Palamunai, nearly 70 per cent of the holdings were under the old *podvarvelian* system of tenancy, the landowners being commercial and landowning people from neighbouring villages.

Finally, in Palannoruwa more than half the cultivators are tenants, operating 27 per cent of the land. This study compares the tenant-landlord arrangements before 1960, and at the time of the study in 1971. Important improvements followed the introduction of H4, a locally-developed rice variety, pushing yields from 16 bushels to about 30 per acre. Net returns for tenants are shown as rising from Rs. 73 to Rs. 142 without taking into account tenant family labour, and for landlords from Rs. 13 to Rs. 117.

Owing to the rising costs of consumer goods and livelihood necessities, the increase in income was less dramatic than appeared. Nevertheless, the author of this study (*Selvadurai* in *Hameed, 1977*) insists that positive improvements in living standards were visible for all categories except for agricultural labourers without land, with the most noticeable gains being those affecting the better-off villagers. But they report tension and conflict between tenant and landlord over the cost of inputs. Landlords continued to claim their half of the total output but were unwilling to go beyond their customary contribution of seed and a limited amount of fertilizer, leaving the tenant to finance the additional inputs required by the new technology. Nor were they willing to take into account the increase in the cost of wages and of draught animals hired. Insecticides also had to be supplied by the tenant.

Selvadurai also observes that the unrest provoked by the failure of the landlords to participate in the increased costs of the new technology has been aggravated further 'by the open flouting of the 1958 Paddy Lands Act', the purpose of which was to safeguard the tenant. The 1970/1 Register contains the names of only 17 tenants whereas their actual number was estimated by field workers as between 75 and 100. This left the majority of the tenants no written proof of their tenurial status and consequently no way of claiming the legally specififed three-quarters share of the produce for meeting the additional costs of production and enjoying improved livelihood. (*Hameed, 1977*, pp. 201–5.)

(ii) Java

But of all the Asian countries, the struggle for land is most acute, most desperate, in Indonesia, or rather in Java. Under the pressure of rising population, tenancy began to spread rapidly from 1900 onwards and various sample surveys done in 1945 indicate that by then at least 40 per cent of the island's families had no paddy lands though they probably owned house lots, while the agricultural lands owned by another 40 per cent of rural families fell short of the livelihood threshold, which was considered to be 0.5 hectares of irrigated paddy land. The 1963 Agricultural Census gave the figures for land ownership as shown in Table 11.

Palmer (1977, p. 131) estimates that 50 per cent of the rural families of Java and Madura are at the present moment landless. But however unreliable figures for land ownership may be, it is even more difficult to estimate the extent of

TABLE 11

Land Ownership in Java—1963 Agricultural Census

Size category (in hectares)	No. of operational farms (in millions)
of less than 0.1	2.2
0.1 to 0.49	4.15
0.5 to 0.99	2.16
1.0 to 4.99	1.16
5 and over	0.034

tenancy. Under the pressures of a struggle for sheer survival, tenancy and debt have become woven together in a fabric of extortion and dependence that is difficult to unravel. There is great variation in forms of actual tenancy, which are easily confounded with forms of pledging land against a loan and of having to make it available to a creditor in lieu of debt repayment. When this happens, the true owner may become a tenant on his own land, with rights to two-sevenths to one-half of the product, according to the amount and type of indebtedness. This kind of situation makes it difficult to distinguish ownership and tenancy.

The most usual form of contemporary tenancy (*Palmer, 1977*, p. 129) is known as *mrapat*, by which the tenant receives one-quarter of the crop, providing only his own labour, so that he becomes a type of share-labourer. Such is the desperate competition for work that a further turn of the screw can be added by the practice of *sromo*, which involves the payment of a premium to the landowner for the privilege of renting his land on these extortionate terms.

Not only is tenancy confounded with the pledging of land for debt, but the share-tenant is forced into a situation where his labour becomes a cheaper option for the landowner than a hired labourer, while for the tenant there is the advantage of at least being able to rely on having work throughout the season.

Wertheim, in a personal communication, reports that the share-cropping law of 1960 was more or less enforced by the

Peasant Unions between 1960 and 1965, but that since the suppression of the Unions it has been completely ignored.

(iii) West Malaysia

Tenancy also plays a large part in West Malaysia, where only 40 per cent of cultivators are full owners of their land, 25 being part owners and part tenants, and the remaining 27 per cent being tenants with no lands of their own. In Permatang Bogak, the village studied for Global Two, twenty-four of the cultivators were found to be tenants with only nine full owners and eleven owner/tenants. To what extent the official over-all figures concealed a higher proportion of tenancy, as seems to have been the case in Sri Lanka and India, we are unable to say.

In the village studied, four-fifths of the tenants adopted the HYV of rice known as *bahagia* when it was promoted by the Department of Agriculture. However, in 1971, all adopters were caught in a sharp cost/price squeeze and three-quarters of the tenants who had adopted the *bahagia* rice decided to return to cultivating the *mahsuri* variety, which fetched a higher price on the market and cost substantially more to produce. Bhati (1976, pp. 147–150) makes the following comments:

Tenants, in spite of the small size of their farms, their complete dependence on rented in land and their low total net income (both in comparison with the owner and owner/tenant cultivators of the village) are fairly capable and innovative farmers . . . but they cannot afford to take big risks as the other farmers can. Every season tenant farmers have to part with large sums for their rent which leaves them with low incomes and a very low level of livelihood.

A further change to be seen in Malaysian tenancy seems to underline the important point that the polarizing effects of the Green Revolution, as between cultivator and controller of land/capital, is not due to technological factors but to the dynamic effect of the potential high-profitability reward for investment. In this case, the proximity of the area cultivated in foodgrain to the city boosts 'mercantility' and the attraction of profitable landlordship becomes sharper.

Bigger Burden on Tenants

An interesting account of tenancy in transition is given for this area (*Duewell and Noor, 1971*, Ch. IV–VI), where, under customary law, a cultivator gained rights to a piece of land by virtue of his occupation, clearance, and utilization of such land. The system of usufructuary rights that accompanied shifting agriculture generated proprietary rights as more permanent local government developed; and in time landed property could be inherited, acquired by marriage, or sold.

The authors point out that share-tenancy in its earlier form 'served a social allocating function that helped to maintain economically efficient and manageable levels of paddy holding'. The landlord–tenant relationship had not been one of domination or class difference and the tenancy arrangement had obliged the proprietor to participate in the harvest labour, prior to taking away his share.

Changes in the institution of share-tenancy observed and analysed by the authors in the riverine areas contiguous to Kuala Trengganu are a response to a number of distinct but related factors. The fertility of the area and the availability of water have fostered settlement and population growth, leading to pressure upon land and tendency towards fragmentation. At the same time, the development of urban population centres and the growth of an efficient system of transport between the urban and the rural areas have increased the value and price of land.

'Tenancy', say the authors, 'is no longer controlled by the operation of laws of socio-economic consensus on a village level, but rather by the demands of the market. In this process, bargaining power is in the hands of the landlord, and the tenant is forced to assume more of the burden and costs of production.'

These changes in agrarian structure are accompanied by a deterioration in the socially-oriented control over the landlord–tenant relationship hitherto exercised by the village community and the face-to-face relations between the two parties as fellow cultivators. One of the results of the movement of landowners to the towns is the depersonalization of landlord–tenant relations, which now respond essentially to the flux of supply and demand rather than the exigencies of livelihood-seeking in a situation where family livelihood needs are approximately equal, but where land endowment is unequal.

The conclusion arrived at is simply that 'the greater the degree of land scarcity with its accompanying socio-economic conditions, the greater the cost-burden shift to the tenant'. It is suggested by the authors that the reduction of the landlord's obligation at the expense of the tenant 'lessens the chances for agricultural and economic growth in the region'.

What their figures most successfully show is that the level of livelihood of tenants is lowest in the area of highest mercantility and land value. The incidence of tenancy is highest in the richest areas, where irrigation is effective, and where there are favourable market conditions and adequate transport.

The introduction of improved seeds, the use of increased quantities of chemical fertilizer and the practice of double-cropping, and the changing terms of the agreement between landlord and tenant in the areas where the value of land is highest, all are best illustrated by following the change in Kuala Trengganu South, where land reached its highest value. With regard to the introduction of chemical fertilizer, the authors note an inverse relation between landlord participation in fertilizer cost and the value of the land, with the tenants having to provide 100 per cent of the fertilizer costs in the case of Kuala Trengganu South.

There was no change in the division of responsibility for provision of seeds and land preparation and tillage up to harvest time since both these were the responsibility of the tenant. However, changes were observed in the Kuala Trengganu South area in relation to the landlord's participation in the harvesting. Hitherto, it had been the practice for the cropped land to be divided before harvest, leaving to the landlord the responsibility of harvesting his own half of the field and carting the paddy to his house or granary. However, in 25 per cent of the cases studied in Kuala Trengganu South I and II and 13 per cent of the Kuala Trengganu North region (i.e., the double-cropping areas) the tenant had to bear the cost of harvesting the landlord's share and thus had to support the full production costs.

A similar burden-shifting was noted in the transport of

the landlord's share, which had hitherto been the responsibility of the landlords themselves. But in the regions near Kuala Trengganu where, in addition to having the most valuable land, the highest proportion of the landlords are absentees, in only 16 per cent of the cases did the landlords transport their own grain to their houses.

VIII   EFFECTS OF MECHANIZATION

The different impacts of mechanized irrigation, the tractor, and the combine harvester are examined, especially the effect on jobs and livelihood.

Cereals cultivated by means of the new technology require an increased input of power whether this is supplied by man or machine, as well as increased inputs of energy in other forms. Changes in husbandry involve the preparation of the land, the application of chemical fertilizers both before planting and as a top dressing during growth, the treatment of seed with chemicals for protection from destruction and to ensure germination, the timing of transplanting, the density of the plant on the ground, the depth of the seed in the soil, the pattern of row-planting to facilitate weeding, the correct timing of irrigation and its control, greater attention to weeding on account of heavy fertilization, and the protection of the plant against pests and disease by spraying and fumigation. All these practices require more than the accustomed amount of work.

A good example of the alternatives and the problems connected with choosing between them is given by Hameed (1977) from Sri Lanka. The customary method of controlling the growth of weeds there consisted of the impounding of water in the paddy fields, which prevented their growth. The greatly increased use of fertilizers substantially increased the luxuriance of weed growth, but it was found that the prolonged presence of stagnant water in the paddy fields diminished the effects of the fertilizer on plant growth. Alternative methods adopted were, on the one hand, the increased use of manpower for hand weeding and, on the other, the capital-intensive spraying of imported herbicides.[1] A third method, practised widely in the Philippines, consisted of the use of

---

[1] A labour-intensive variant reported as practised in the field without official sponsorship consisted of dabbing the emergent weeds individually rather than spraying them.

light power-driven hand-operated weeders. Clearly each of these involved an increased energy input in the process of production.

Changes in husbandry have far-reaching results in themselves that are the more spectacular where agriculture is predominantly self-provisioning and rudimentary. Their consequences lead in several directions: the operations of agricultural production appear to become more complicated and demanding, requiring increased manual labour or else its substitution by machinery or chemical means. The use of chemicals requires expenditure in the market, which implies the use of credit, and a commitment of the small economy to the market.

It is clear, however, that cultivators would not encumber themselves with more complicated methods of production and heavier costs but for the prospect of significantly increased yields providing for improvement in the security and fullness of livelihood or giving them enhanced net gains from the productive activity.

The way in which technological elaboration pushes the cutivator into making new choices between economic means is well illustrated in a passage from our report on rice in Monsoon Asia (*Palmer, 1976*):

It has been noticed in the Philippines that farmers planting HYVs put more effort into ploughing and harrowing in order to incorporate weeds and straw more thoroughly into the soil. This has encouraged the mechanization of land preparation rather than greater human or animal effort because it has been found to be more difficult to obtain the custom service of buffaloes and ploughers and they are expensive to hire.

An FAO report of the Philippines HYV programme (*FAO, 1971*, II) tells us that 'cost comparisons on the basis of current hire rates of hand tractors and buffaloes for land preparation show that the cost of ploughing one hectare with a one-axle tractor is only slightly higher than with *carabaos* (buffaloes), and that harrowing is substantially cheaper with a tractor because the work is done in a much shorter time'.

The repercussions of the increased work/energy requirements of the new technology and the complexities of the choices facing the cultivators are further demonstrated in in connection with the synchronization of the planting of HYVs with other varieties. Part of the new technology packet introduced in the Philippines was the highly product-ive cultural practice of transplantation, and this made serious demands on labour. Improved yields also amplified the labour requirements of harvesting and threshing. Both of these operations threatened to produce a peak labour demand period but for the fact that the HYVs were quick-maturing, and if planted at the same time as the local varieties, would be harvested three to four weeks before them, thus confining the peak to the planting season. Or else, the planting could be delayed so that the harvests were synchronized.

However, two other considerations had to be weighed in the choice: if planting were staggered, the exploding popula-tion of pests infesting the maturing plants would quickly transfer its attention to newer shoots growing up in adjacent lots, and would be likely to do disproportionate damage. The second consideration to be weighed in deciding when to plant was the advantage of getting maximum benefit from the early showers of rain. Should the decision favour synchronized planting, the peak demand for labour was met by bringing in women and youths to form transplantation teams. But if the decision favoured staggered planting, then the synchronized harvesting of increased volumes of grain was likely to resolve itself sooner or later in mechanization (*Palmer, 1976*).

Thus the intensification of tasks involved decisions in-evitably leading to changes in the economic and social fields. And the elements intervening in the choice were so various as to make prediction very difficult.

Radical Changes

Before getting into the effects of mechanization, it is interest-ing to see that the promise of profitability and the market involvement brought by the high-yielding varieties was able to lead to radical changes at a very rudimentary level in Java. We are referring to the appearance of the *tebasan* system of contracting out the harvest and to the substitution of the sickle for the *ani-ani*.[2]

---

[2] The *ani-ani* is a harvesting knife, and as each ear must be severed separately

Prior to these innovations, harvesting was carried out in a customary manner within villages in such a way that villagers could expect to receive small stores of grain in return for their participation, and also in payment for certain labour services rendered by them during cultivation for which they had not been paid at the time. Each harvester was paid an agreed proportion of the paddy harvested, which varied according to degree of kinship of the harvester to the cultivator. Latterly, in response to the increasing rural population and the dwindling availability of grain per capita, harvests became crowded occasions for the poor not only of the village but of the neighbourhood. It became difficult for the cultivator to control the actual size of the paddy bundles carried away by each harvester, and this circumstance threatened his ability to pay off his other production costs out of the sale of his anticipated harvest (*Collier, Soentoro, and Wiradi, 1973*).

TABLE 12

Average Customary Share and Tebasan Share of Harvested Crop: Dry Season, Central Java 1972

Village	Customary Share		Tebasan
	Custom	*Actual*	
Village No. 1, Pemalang			
Average share	1:8	1:6	1:11
(Number of observations)	(30)		(23)
Village No. 2, Pemalang			
Average share	1:9	1:6	1:12
(Number of observations)	(29)		(15)
Village No. 3, Kendal			
Average share	1:9	1:6	1:11
(Number of observations)	(21)		(3)
Village No. 4, Kendal			
Average share	1:7		1:9
(Number of observations)	(24)		(2)

Source: Collier, Soentoro, and Wiradi, 1973.

it makes harvesting a slow process. It also has religious significance. In comparison with the *ani-ani*, the sickle represents an important technical advance.

Under the *tebasan* system, the cultivator sold his standing crop to a contractor (*penabas*), who brought in his harvest team, usually paid in cash, and received a proportion of the harvest that was smaller than that owed by custom to the harvesters. Members of the Indonesian agro-economic survey studied the new system in a number of villages and witnessed its obvious advantages for the cultivator (see Table 12).

The economic issues involved are seen even more clearly in Table 13. The social repercussions of this critical change are discussed later in this chapter.

Bartsch (1972) has made a useful contribution to systematizing the discussion of the employment effects of technological change. He distinguishes between 'technologies'

TABLE 13

Harvesting Costs and Returns under Three Harvesting Conditions in a Village in Kendal, Java, Dry Season 1972

	Ani-ani plus customary shares of 1 and 6	*Ani-ani* plus *tebasan* shares of 1 and 11	Sickle plus payment in money
Number of harvesters			
Average per hectare	184	150	80
(Number of observations)	(22)	(2)	(3)
Gross yield PB rice			
Average ton of paddy per ha.	4.72	4.72	4.72
(Number of observations)	(7)	(7)	(7)
Cost of harvesting			
Percentage share to the harvesters	14.3	8.3	*
Cost in rice (ton of paddy per hectare)	0.67	0.39	*
Real or imputed** money cost in rupiahs per hectare	10,050	5,850	7,560
Return to harvesters			
(Rupiahs per hectare)	55	39	95

*When sickles are used no payment in kind is made.
**Based on the price during August harvest time of 1,500 rupiahs per quintal of dry stalk paddy.
Source: Collier, Soentoro, Wiradi, 1973.

('particular combinations of material inputs of a biological/ chemical nature for producing a crop') and 'techniques' ('methods of delivering inputs in the productive process, i.e. types of power sources and associated equipment') and gives to each a three-stage rating, thus:

Technologies may be:	*Techniques may be*:
traditional (traditional seeds, rainfall or traditional irrigation, no chemical fertilizers or agricultural chemicals)	*traditional* (manual, or manual plus animal power)
improved (substitution of one or more 'improved' inputs for traditional ones)	*intermediate* (manual plus animal power)
modern ('package of high-yielding seeds, chemical fertilizers and controlled irrigation; plus agricultural chemicals')	*mechanized* (power-driven machinery with associated equipment)

Taking a first transition from 'traditional' to the new technology, (which he calls 'modern'), and basing himself on seven case studies of production practices for wheat, rice and *bajra* (a low-quality cereal), he finds that the labour input per hectare is considerably higher under the new technology with traditional techniques, while it is 23 per cent lower per unit of output, and 48 per cent lower (per unit of land) where improved techniques ('mechanization') have been adopted along with the new technology.

When he compares a bullock-powered technique with a partially mechanized one, he finds that in the four studies available for analysis, there are reductions in the labour input per hectare of between 3 per cent and 37 per cent.

To gain a clearer picture of the effects that 'intermediate' techniques and full mechanization are likely to have on employment, the author has built up hypothetical models, based on Indian employment data of the three techniques for wheat production in India. These models indicate that man-hours per hectare are reduced 74 per cent and 90 per cent respectively when moving from 'traditional' to 'intermediate' and 'mechanized' techniques under 'traditional' technology and reduced 77 per cent and 93 per cent respectively if the technology is already 'modern'.

Thus, it appears (Bartsch says in conclusion) that both 'intermediate' and 'mechanized' techniques are very labour-saving, but especially the latter.

Using these Indian models, Bartsch also compares the employment effects under various combinations of technology. He concludes that the combination of modern technology and fully 'mechanized' techniques utilize only about 10 per cent of the labour used by a combination of 'traditional' technology and 'traditional' techniques. And when 'modern' technology is combined with 'intermediate' techniques, labour input per hectare appears to decline to about 40 per cent of the traditional technique/technology. Therefore the labour-displacing effects of using either 'intermediate' or fully mechanized techniques more than offset the labour-augmenting effects of using 'modern' technology.

The main conclusion is that the only way to avoid a reduction of labour input per unit of land and per unit of output is by adopting the new 'modern' technology but retaining traditional techniques, in which case labour requirements can be increased. Broadly speaking, this conclusion is reinforced by field observations, but each situation manifests its own idiosyncracies.

The tractor's main uses are for land preparation, threshing, and transport. It has a specially important function to perform where soils are hard and dry, as in Morocco and Tunisia, or for dry season cultivation in Asia. The tractor can also get land ploughed much faster than the commonly used draught animals can, and this is also a major consideration when ploughing must be completed between the falling of the first showers and the coming of tropical torrents. Its contribution to the speed of operations may also open up entirely new multi-cropping possibilities, provided water is available throughout the year.

In many of the areas in which the new technology was introduced, tractors were already being used by the majority of cultivators, e.g. in North-West Mexico, Surinam, West Godavari, West Pakistan, and Minipe, Sri Lanka. In others, tractor use increased greatly after the introduction of the new technology; for instance in Maraliwala, Pakistan (*Chaudhari and Rashid, 1972*), use rose from 46 per cent to 91 per cent of the farmers; in the Malaysian village studied by U. N. Bhati (1976) all 44 farmers became users, while the number of tractors owned by the sample farmers of Gurdaspur (*Kahlon and Singh, 1973, I*) doubled from 37 to 74 in the four following years.

Tractors Take Over

Barker (1972) has succinctly put together the relevant facts on the employment effects of the new technology for the Philippines. Between 1960 and 1970, the area under paddy and the total amount of labour used have remained constant, while production has risen, and with it the productivity of labour, from 34 to 25 man-days per ton. At first sight, this appears to be a measure of the contribution of the new technology and demonstrates its land-saving effect. But it must be observed that during the same period the productivity of labour rose almost as much on farms generally, suggesting displacement of labour by tractor power in the cultivation of local varieties as well.

Barker's research showed the following:

Land preparation   A reduction from 17 to 10 man-days was effected as a result of partial mechanization. During the period the proportion of cultivators using tractors (mainly custom-hirers) rose from 14 to 49 per cent. It is noted that the most common practice was to have the first ploughing done by hired tractor but then to use the water buffalo for harrowing.

Transplanting   No change.

Weeding   The heavy crop of weeds nourished by fertilizers required the doubling of labour used in spite of the increasing use of mechanized weeders and of herbicides.

Harvesting and   The amount of labour increased with
threshing   yields but there was little change in tech-
                     nique. Most cultivators continue to use
                     the mechanical threshing machine, intro-
                     duced long before the HYVs.

The net position emerging from these calculations suggests
that the annual increments in production required to maintain self-sufficiency in rice can be met by a continuing
improvement in both land and labour productivity through
improved irrigation, and more effective application of the
new technology. This would not make allowance for increasing the land and occupation opportunities for the annual
incremental labour force, and particularly for the sons of
tenants.

Were cultivators not tenants, no doubt land fragmentation
would have taken place long ago. As it is, while one of the
tenant's sons may expect to take over his father's tenancy,
the remainder have proletarian prospects. Up to a certain
point, proprietors have been willing to allow sons, if they
remain in the locality, to build houses on their fathers'
rented land, but the limit is soon reached. In view of the
scarcity of job opportunities, the uncertainty of livelihood
and residence amongst the sons of the tenants is an acute
source of tension.

Barker implies that continued labour absorption could be
expected if government policy refrained from subsidizing the
use of tractors—a conclusion that tallies with those of Bartsch.

Freebairn (in *Poleman and Freebairn, 1973*, pp. 107–9)
also distinguishes between the effects of mechanization on
the various operations. He takes the position that several
factors work to make the modernized agrarian system capital
intensive, but the displacement of labourers by more mechanized production systems is offset by multiple-cropping and
by the jobs that may be generated to service mechanized
agriculture.

A study from the Punjab, India, used as a basis for discussion, shows the effects on employment when mechanization
is introduced. Farming operations are divided into four
categories in order to evaluate the potential for mechanization

and the probable influences on employment. The categories used are: (i) seed bed preparation; (ii) irrigation; (iii) inter-culture; (iv) harvesting. The study further distinguishes between effects on family labour, permanently employed labour, and daily wage labourers.

Freebairn draws the following conclusions:

If mechanization is adopted, it may reduce the total labour requirements by 20–30 per cent. But land preparation, seeding, and irrigation are jobs usually done by family labour. If machinery takes over, no labourers' income stream is diverted and indeed family livelihood could be considered to have been improved. The tractor is not likely to contribute to interculture, which continues to be labour intensive. However, he found that in 1969, one-half of the wheat harvest of the Punjab was already being mechanically threshed and he predicted that by 1974 all would be. This, he estimated, would mop up 90 million man-days per year, hitherto earned by wage labourers.

Special attention was given to the role of the tractor in Sri Lanka by our study there (*Hameed, 1977*).[3] More than 12,000 four-wheel tractors and 3,500 two-wheel power tillers were reported to be in the country in 1969.[4] Table 14 shows the proportion of farm units and paddy areas in which tractors are being used in the various districts.

It will be seen from the table that in six out of twenty-two districts in the country, more than 50 per cent of the area under paddy is ploughed only with tractors. When we examine this with reference to observations in the localities of our study, we find that tractors are being commonly used in three out of four localities studied, for land preparation, threshing of harvested paddy, and transportation or haulage purposes. In the Minipe area, where the degree of mechanization is relatively high, there are 87 four-wheel tractors and 56 two-wheel power tillers. These figures represent ownership by 11 per cent and 5.5 per cent respectively of the cultivators. The average distribution of this machinery here has been estimated at one tractor for 52 acres of paddy land.

[3] This account was written by Abdul Hameed and appears also in the Sri Lankan report.

[4] See *Raj, 1972*.

TABLE 14

*Proportion of Farm Units and Paddy Areas using Tractors
in the Various Districts*

	Farm units using tractors only	Farm units using tractors and animals	Area of land using tractors only	Area of land using tractors and animals
Amparai	84.4	7.3	73.5	6.9
Anuradhapura	15.7	10.3	27.7	8.2
Batticaloa	59.8	1.9	67.3	4.3
Colombo	5.1	1.3	8.6	1.3
Galle	0.8	0.5	0.9	0.8
Hambantota	73.3	3.1	80.5	3.6
Jaffna	33.6	24.1	49.5	50.5
Kalutara	0.1	0.2	0.08	0.3
Kandy	0.07	0.2	0.1	0.3
Kurungala	21.5	10.0	23.9	12.8
Mannar	19.3	63.2	16.9	67.7
Matale	14.5	2.4	13.1	2.6
Matara	12.7	2.0	18.4	2.0
Moneragala	30.6	16.3	24.8	19.6
Nuwara Eliya	0.05	0.6	0.07	0.8
Palannoruwa	7.4	25.9	7.1	28.2
Puttalam	41.7	8.3	50.2	14.7
Ratnapura	3.1	–	8.2	–
Trincomalee	53.2	6.3	60.3	5.8
Vavuniya	68.7	31.3	70.2	28.5

Source: Survey on Cost of Production of Paddy, Central Bank of Ceylon, Department of Economic Research, Colombo, 1969.

Most of these tractors are owned by the private farmers in this locality and hired out at the rate of Rs. 55 to 70 per acre for ploughing and Rs. 30 to 40 for threshing operations. Of the farmers studied in this locality, 95 per cent are using tractors for ploughing, 82 per cent for threshing and about 30 per cent for transportation purposes. (*Amerasinghe* in *Hameed, 1977.*)

Similarly, in Palamunai village (East Coast), every one of the households studied has been found to use tractors for preparatory tillage and threshing operations, although here the proportion of farmers using tractors for haulage purposes is only 4 per cent of the total farmers studied (Selvanayagam in *Hameed, 1977*). Unlike those in the Minipe area, most of the tractors used in this village are owned by wealthy landlords (*podiyars*) living in the neighbouring villages. In the third locality, Palannoruwa (Western Province), 40 per cent of the farm households studied use tractors, 22.5 per cent use both tractors and work animals and the remaining 37 per cent use only work animals for ploughing purposes.

As regards threshing, 57.5 per cent of the farmers in this locality are reported to use tractors. Most of the tractors now used in this village are owned by high-income groups, living in this area (Selvanayagam, in *Hameed, 1977*). In the fourth locality, Ussapitiya village (Kegalle District), farmers do not use tractors for rice cultivation.

Motives for Mechanization

It should be stated at the outset that in Sri Lanka the use of tractors has not been necessitated by the introduction of the new biological and chemical inputs. In fact, the use of tractors for cultivation purposes appears to have started during the early 1950s, about ten years before the large-scale diffusion of HYVs. The evidence yielded by our locality studies suggests that tractors have not induced farmers to use new inputs or adopt new cultural practices, nor have they resulted in a notable increase in the incidence of double cropping. On the other hand, the use of tractors appears to be having a number of consequences with regard to the use and organization of local resources, including labour, and farm costs, which in turn have a bearing on the national problems of employment and foreign exchange.

As a consequence of the widespread use of tractors for land preparation, threshing, and haulage operations, there has been a decline in the amount of farm labour previously used for these operations. It has, on the other hand, increased the demand for the services of certain skilled non-farm labourers (such as tractor drivers) and non-farm inputs (such as fuel). In most areas, farm costs are reported to have risen due to sharp increases in the hire charges for tractors. In many other cases, farmers are also reported to be facing the problem of non-availability of tractor services in time—a problem in turn attributed to the non-availability of foreign exchange for importing spare parts—which disrupts the cropping routine.

Why people want to own and use tractors in Sri Lanka is

a question worth examining here. The results of a recent investigation into the motives of tractor owners for buying this machinery, and the reasons given by farmers for shifting to the use of tractors, are helpful in explaining the benefits that the individual owners and others expect and get from the use of tractors. According to this study (*Carr, 1975*), in the case of four-wheel tractor owners the use of tractors 'for transport and haulage and the increased income to be made from custom work' is highly emphasized. In the case of farmers who hire the services of tractors, the study reports that their 'motivation seems to be to relieve the problems involved in hiring labour, and in either hiring or maintaining buffaloes. The thought of increasing returns per acre by more timely operations also seems to be more important to this group than to the tractor owners.'

As regards the benefits achieved by changing over to tractors, the study referred to here has indicated that the 'beneficiaries of tractorization are those tractor owners who are relatively well off to start with, such as businessmen, plantation owners, large landowners, and village entrepreneurs. These people in fact own a major proportion of all tractors and have an almost total monopoly of the new, more efficient tractors'.[5]

As for the users of hired tractors, the study has reported that they 'realize few benefits because of the inefficiency and increasingly high price of custom services. . . . The major benefits reported by the users are increased leisure and better working conditions'. According to the study, 'many farmers reported that these compensated for the higher costs involved in using a hired tractor rather than their own animal power'.

In other words, the general tendency on the part of the farmers is to prefer the use of machines in place of work animals, despite the fact that the former costs more than the latter. This is supported by the results of Global Two studies in Minipe and Palamunai. According to the author of the Minipe study, the availability of buffaloes per lowland paddy acre was higher in Minipe, in comparison with the other

---

[5] Minipe, a recent area of government-sponsored and assisted settlement, is an exception to this generalization.

special projects, in 1967–8, and the buffalo to tractor expense was 1 : 4; yet, farmers preferred tractor ploughing. (*Hameed, 1977.*)

In official circles, the prevailing view is that the large-scale import and use of tractors is justified by the shortage of work animals and human labour during peak seasons on agricultural activity. In a country like Sri Lanka, which depends on imports for all its tractor requirements, spare parts, and fuel for machinery, and which faces the problems of declining foreign exchange earnings and growing unemployment, one might expect that this official attitude would not continue for long.

One of the underlying causes of the scarcity of animal and human labour felt during the peak seasons is that land preparation, threshing, and transporting operations are concentrated in certain periods of the year. This in turn is explainable in terms of the commencement of the agricultural calendar immediately after the onset of the first rains and uniformity in the timetable of rice crops grown in most parts of the country. If appropriate measures, including an improvement in irrigation facilities and cultivation techniques and the introduction of varieties characterized by different maturing periods, could be taken, there would be a possibility of relieving this scarcity without resorting to extensive use of tractors.

It is the combine harvester that brings the most sweeping reductions in labour-use to cereal production, and it is the difficulty of finding an efficient system of mechanized harvesting that has kept labour needs for cotton and sugar cane so high. In 1972, there were said to be twenty combine harvesters in the Indian Punjab. A number were also introduced experimentally into Malaysia. Since they are capable of doing the work of ninety labourers in a single season, and since harvest work is frequently more highly paid on account of the conditions of peak demand, any generalized use of combine harvesters in a labour surplus country would have catastrophic results for rural livelihoods.

Mechanization in the new technology areas of Asia is primarily represented by the tractor and the tubewell. However, policy or no policy, it is exceedingly difficult to find examples of government action that have been capable of arresting the push towards labour-saving technology where this is shown to be highly productive and profitable to the entrepreneurial class. It is with this in mind that reports on the situation in North-West Mexico are introduced as an example of a 'bi-modal strategy', where agriculture is divided into a commercial sector and a subsistence sector. (Compare Barker (1972) on labour absorption in the Philippines.)

PART III

IX THE CRITICAL ISSUES

*The findings of the study are summarized: in unequal
societies the new technology can facilitate 'take-off' for
cultivators with land and some capital but institutes
changes that marginalize the small cultivators without
capital and land and undermine the essential and customary means of livelihood of an ever-increasing number
of people in rural areas.*

The purpose of the present chapter is to draw together the
critical issues that have come to light or been verified by the
UNRISD studies of rural localities and their response to
programmes for the introduction of the new technology. This
leads us to re-examine the intrinsic consequences of the
technology, the appropriateness of its introduction at a particular place and time, and the strategy embodied in the
programmes and other measures by which the introduction is
effected.

The re-examination is made more difficult by the fact that
the introduction of a technology on a large scale is an inter-
vention in a complex situation that includes social forces
with a potential for dynamic change, and that outcomes
depend on differences in agrarian structure and on the degree
of industrial development and of infrastructural elaboration.
They also depend on the capacity of governments to establish
efficient services, on the style of development pursued by the
government, and on the extent to which it enjoys or looks
for political support from one or another of the classes
engaged in the productive process.

I. THE ESSENTIAL CHANGE SEQUENCE

Underlying the differences between programmes and between the social situations in which they operate, there is a common change sequence issuing from the intrinsic nature of the technology and its immediate results:

High-Yielding Varieties (HYVs) require a greater input of energy per unit of land—especially of nitrogen and of motive force—for a more complex, more controlled husbandry than is required in customary cultivation routines, in which local varieties are used that rely heavily upon their environment and are adapted to its idiosyncrasies. Amounts of nitrogen in the quantity required by the recommended packages are such that in most cases this input can be supplied only by manufactured artificial fertilizer. Irrigation requirements are best satisfied by canal systems or by ground water, and in practice controlled supplies of ground water, requiring oil-driven or electric pumps, have been the most effective suppliers of controlled moisture. In most of the areas under consideration, efficient irrigation is therefore also a consumer of energy.

At the same time, where the technology is agronomically successful, considerable increases in yield follow, producing an even greater rise in the market surplus.

The great increase in the surplus of grain offered for sale and the similarly increased purchase of industrially-produced inputs and means of production combine to increase greatly the *mercantility* of the farm or unit of production.

In thus changing the agronomic and economic character of the farm, the optimal requirements for the success of a cultivator are also changed. Where it is possible to combine economies of scale offered by labour-saving machinery with a strong bargaining position in commercial transactions and access to cheap capital, a high level of profitability may be achieved. In areas favourable to the new technology, therefore, farming becomes an attractive investment and novel conditions prevail, changing many existing relationships.

II. INCORPORATION AND EXTERNAL DEPENDENCE

Wherever the new technology replaces or is added to the

older agricultural systems on a large scale, and mercantility inevitably increases, there is further incorporation of local economic systems and livelihood support patterns in the urban-industrial macrocosm. The process of incorporation has different consequences for different rural strata.

Thus, local cultivators must purchase fertilizers, chemical products, machinery, fuel, and machine maintenance from the industrial sector: their seeds, propagated by scientific research centres, are obtained through urban distributors or large-scale cultivators, usually from outside the locality. Local cultivators also come to depend upon technical services and suppliers of institutional credit from outside. They must learn to sort out, decode, and evaluate the scientific and economic messages that reach them from bureaucracies, banks, and experimental stations.

The implications of increased dependence of the locality upon the urban-industrial network is an aspect of social change that leads in several directions. It biases the distribution of advantage in favour of those who have the experience and social attributes necessary for confronting the city and the bureaucracy, the printed instructions, and the political caucuses; and it puts a relative handicap on those whose assets include traditional knowledge of the local idiosyncracies of soil and climate,[1] and whose energies are absorbed by the labours of husbandry rather than in manipulating the rural–urban nexus.

External dependence implies a swing away from local self-reliance; it implies the local community and the individual productive unit becoming a part of a larger system of production and exchange that has a potential for diversifying and enriching life and livelihood. In the agrarian society with a low technological level, most rural families must produce the food they live by, and in this sense they enjoy some security, though subject to the chances of regional catastrophe and local extortion. The linkage established by the new technology between the local community of producers and the larger society tends to withdraw much of the decision-making

---

[1] The weight and prestige given to the new expertise is quite capable of overriding local experience in such a way that practice suffers.

autonomy from the former, and subjects it to national and international episodes of politics and trade and the repercussions of distant war.

Self-provisioning is based on attachment to a particular piece of productive land, and it is the ultimate refuge of the peasant, enabling him to opt out of the confrontations and struggle. To embrace the new technology and the commercialization that accompanies it may increase peasant incomes, but implies a movement away from this refuge. Colonial history offers many examples of market flux, now leading peasantries into market production, now leaving them to readapt to fuller self-provisioning. While the large farmer with resources and investment alternatives may deliver himself entirely to capital-intensive market production of cereals, so long as this is profitable, the poor cultivator with a tenacious hold on his land may attempt to retain a self-provisioning capacity as free as possible from debt.

The new external dependence of the locality also makes for changes in its power structure. Power based on hereditary land monopoly is rivalled by power accruing to those who can control the traffic between the local community and the larger society, the 'nexus people'.[2]

This process of incorporation, of which the propagation of the new technology and the majority of rural development programmes both form a part, places many of the blessings of science and industry within the range of vision of rural people, though their capacity to take advantage of the offerings depends on their assets, their socio-economic status, their credit-worthiness.

In order to appreciate the social and economic implications of the new technology, therefore, it is necessary to look at its consequences in the framework of the larger process of

---

[2] Namely, those who manage commerce and the officials, whether outsiders or insiders, who manage government agencies connected with law and order, health, public works, agriculture, and development programmes; and those who become the recognized political chiefs, and are responsible for arousing and maintaining support for their parties among the local population, and canalizing those favours that the parties, in or out of power, can pass down; and those who control communications and transport. All these elements are to be found in the élites who handle agricultural development at the local level in a variety of alliances and compacts.

incorporation, and above all to give some account of terms of incorporation that can be secured by the various categories and classes of person affected—broad categories estabished by their relation to both the old (locality bound) and the new (urban-industrial dependent) productive systems, and their capacity to deploy the assets and attributes they control.

Our explanation, therefore, turns about *the struggle over the terms of incorporation* by different classes of protagonist.

III. EMERGENCE OF THE ENTREPRENEURIAL CULTIVATOR

Entrepreneurial cultivators have appeared significantly in wheat production[3] in Mexico,[4] India, and Pakistan, and have achieved a high level of profitability in relation to units of output as well as on capital invested.

Entrepreneurs have been chiefly owners of middling and large farms favoured by access to capital and access to (and confidence in) technological know-how. This qualification obviously is inclined to accompany a higher level of education, and some familiarity with urban and bureaucratic ways. Moreover, two features of the strategy of most of these programmes have fitted them with a further bias towards polarization and uneven growth, favouring the cultivator in these conditions—namely, concentration of programmes and investment in the best agricultural areas, and the 'progressive farmer' approach.

The first of these elements of policy has much to recommend it both from the point of view of those who wished to mark up rapid and early successes in adoption and yields, and those who wished to build on the highest possible levels of existing investment in infrastructure and productive

[3] Large-scale entrepreneurial production of rice is to be found in many parts of the world, but there is little information available about its emergence in the Asian peasant setting, e.g., in Malaysia or the Philippines. It is, of course, possible to speak about small-scale entrepreneurship in farming. For instance, a Taiwanese cultivator with half an acre of land refrains from subsistence-oriented production in favour of a more profitable crop and freedom to buy and sell—an opportunity that family self-provisioning would not permit. But the Taiwanese paddy cultivator is something of an exception. (See *Wang and Apthorpe, 1974*.)

[4] For an interesting account of the entrepreneurial cultivator as a social category in Mexico, see *Hewitt de Alcántara, 1976*.

equipment. An alternative to this policy would involve a spreading of investment and the application of agricultural and social sciences to the problems of cultivators, especially small ones, working in indifferent and marginal lands as well as those enjoying optimal conditions. Just what it would take to obtain successes by such a policy, both in terms of over-all production results and in terms of raising the farming and living standards of the poorer rural sectors, is discussed more thoroughly in Chapters XI, XII, and XIII. But it would be hard to get such a policy accepted in political and administrative quarters in a society that relied on market forces for its driving impulse—the political will would be lacking.

The other element of policy common to most programmes and tending to accentuate polarization was the so-called 'progressive farmers' approach (see p. 175). It is of course true that enthusiasm for innovation varies among cultivators, and may be found among poor ones. But in practice the 'progressive farmer' commonly turned out to be the cultivator with relatively ample landholdings and access to capital as well as reasonably easy relations with the authorities and an above-average education. It was seldom difficult to persuade such farmers to adopt the package recommendations since they had already made the critical leap into predominantly commercial farming and recognized the deal offered as a good one: inputs, credit, and technical advice were assured and in some cases the product price was supported or at least subject to a guaranteed minimum.

The tactic was widely successful and in most cases the first year's operation, carried out on the best lands of the 'progressive farmers', was rewarded with markedly higher yields. Where successful, the results encouraged other cultivators, many of them less well-endowed with land and capital and less well-connected, to experiment. In optimum areas even poor cultivators strained themselves to obtain credit for inputs. But with the wider diffusion of the HYVs, average yields declined, manifesting great variation on account of the patchy quality of irrigation systems, irregular supplies, increased disease, and improperly applied methods.

Where the 'progressives' were able to demonstrate the unusual profitability of the new technology organized in capitalist farms, cultivation itself as an enterprise began to appear attractive to those with some capital resources. Landlords who had formerly been content to receive share-rents in kind from meagre harvests were tempted to become direct producers, repossessing their rented land from their tenants to work it with hired labour, or to mechanize and to work it with family labour.

Other important consequences, where the new profitability became an established fact, were an increased demand for land and consequently a sharp upward trend in land values, and a vigorous demand for commercial tenancies on a cash rental basis.

Agricultural entrepreneurship also became an attractive option as a sideline for professionals and as a retirement occupation for civil servants, ex-officers of the armed forces, and other middle-class groups who would not have considered it before.

The New Class

In those parts of India where the new technology has shown itself to be patently profitable, and where various forms of assistance have been offered by the government, the effects have been several. On the one hand, landowners have in many cases become direct producers themselves, dismissing their tenants and taking their land under direct cultivation. In some cases, these landowners have been less interested in becoming entrepreneurs but, taking advantage of the widespread demand for access to land, they have changed the terms of tenancies, so that the new profits will accrue mainly to them.

A second trend in India was the return of petty landowners whose social aspiration and capacities had led them to prefer to seek a better livelihood in the towns than could be afforded by small-scale unimproved agriculture.

Joshi (1971, p. 20) points to the increasing importance in India of those landowners 'who are changing over from a feudalistic type of relationship with the direct producers to a commercial and capitalist type of relationship'. He divides these into two groups:

(i) the commercial type who utilizes the traditional agrarian tenancy framework, but 'plays an active part in the supervision and management of land and [is] more inclined to make investments in agricultural improvements for maximizing . . . gains than the old landlord was'; and

(ii) the 'capitalistic type of landlord who has switched over from the use of tenancy to that of wage labour'.

It is the second group whom we refer to as the entrepreneurial cultivator. In the case of the tenant farms on the lands of the first—whatever the aspirations of the tenant cultivators, even if the landowner encourages the tenant to use improved methods, makes certain investments in the means of production, and provides his tenant with credit— the division of entrepreneurial decision-making between the two parties and conflict over the transfer of profits from the tenant to the landlord are likely to hold back progress of this productive unit and the necessary investment in land improvement, irrigation, and productive equipment.

In India, the entrepreneurial farm is mainly in the hands of individual cultivators who have acquired a dominant position in the village as landowners, and who have the resources to operate freely as entrepreneurs, and whose land is of sufficient size to prevent the untimely interference of family subsistence requirements with their freedom to buy and sell. This may mean that he must be able to put by a sufficient amount of grain for household purposes.

Bapna (1973) describes two such entrepreneurial cultivators in the Rajasthan (Kota) study. One of them had been a landowner with tenants and had also worked for the local authorities as a driver. As farming prospects improved with the introduction of HYVs in 1967/8, he took over the cultivation of his own land, amounting to 34 acres, with three permanently employed farm servants. His social position and relationship to the local authorities (*Panchayat Samiti*) gave him a direct relationship to the Block officials, and facilitated his access to technical information and services. As Chairman of the Co-operative Credit Society he was also well placed for receiving institutional credit. He had recently purchased a tractor and a thresher, and with the hiring out of the thresher alone he had earned one-quarter of his total income. He invested substantially in fertilizers and obtained high yields with his wheat, paddy, and pulses. Tractor and thresher ownership gave him a great advantage in speed over cultivators using bullock power, making it possible to grow paddy as a second crop, which was a more profitable crop than *jowar* (a low-quality grain), grown by the bullock-cultivators.

The situation of this cultivator was one of great entre-preneurial resilience and he was free to choose between land-and-water improvement, renting in more land to make better use of his machinery, or the extension of his custom-hire business for agricultural operations and transport.

But there is a second stream flowing into entrepreneurial farming, mentioned by various writers, that consists of 'moneyed men from the business and professional classes, retired members of the bureaucracy and the army, and influential and affluent politicians' (*Joshi, 1971*). It is assumed that the profitability of the new technology has caused many people to choose to become cultivators who otherwise would not have done so. But it is probable that most of them come from landowning families. A large part of the bureaucracy is recruited from peasant proprietors' families, while the military propensities of the Punjabi peasantry are well known.

What is important in India, however, is that this growing class of agricultural entrepreneurs has come to have enhanced political power in its own right during recent years, and already has a powerful voice in State politics on such questions as the price of grain, land ceilings, and the taxation of agri-cultural income.

Above all, the emergent rural middle strata are likely to continue to block redistributive legislation in favour of the land-poor and landless, whose economic improvement they perceive as a threat to their cheap and subjected labour force, their rents, their interest, and their petty monopolies.

In all the situations studied at close quarters, the outlines of this emergent class began to appear, whether as agricultural entrepreneurs (as in India), or as petty landlords and towns-folk whose livelihoods are supported by activities outside the agricultural sector (in Sri Lanka and the Philippines) or as elements of the new bureaucracies (as in certain African countries). Their role in relation to the prospects for small-scale agriculture is a fundamental one, and their different economic functions and modes of operation from one rural society to the next provide an interesting point of entry for understanding agrarian structures.

IV. THE TALENTS–EFFECT AND THE TERMS
OF INCORPORATION

In a landed society, proprietorship is the basis of prestige, power, and the control of other resources. The distribution of land ownership therefore provides a guide to the concentration and exercise of power. Where land is relatively evenly distributed there is little evidence that smallness is accompanied by domination or discrimination. Where there is great inequality in proprietorship, smallness carries with it social handicaps confining entrepreneurial freedom and putting the small cultivator in a situation of contractual inferiority in his market relations as well as in his attempts to obtain legitimate access to public facilities. And where small farms are in tenancies of the common pre-capitalist mode, an additional element of handicap is added that may raise the livelihood threshold to twice its prevailing level for a proprietor in the same locality. It may also involve subjection by permanent manipulated debt.

Our studies revealed that small cultivators lacked the time, influence, literary, and social affinities possessed by the large proprietors that made it possible for the latter to be in touch with government programmes and facilities and receptive to technical information. Thus, peasants may find themselves competitors for credit or irrigation facilities with agriculturists who have city houses and political connections; poor villagers may have to compete for institutional credit with the local élite who make up the village committees that allocate the credit; illiterate, ill-clad cultivators may have to argue their case in town offices with status-conscious officials.

Furthermore, the small cultivators are frequently the dependants of members of the local élites for consumption credit, access to water, use of equipment and facilities, and even for contact with the rural development bureaucracy.

It is the social situation of the small cultivator *vis-à-vis* the purveyors of his inputs, coupled with the economic fragility of his enterprise due to his penurious supply of land, that turns the excellent agronomic potential of the new technology into an indifferent bargain. Continued use of a hardy low-cost technology for his food supplies usually offers him a safer option in the real world, which he knows all too well.

Readers will recall that field studies suggested a surprisingly high proportion of small tenant cultivators in Malaysia, the Philippines, Indonesia, Sri Lanka, and even India, and though the figures that emerged were not reliably representative, the current movement into tenancy involving a steady growth of petty bourgeois non-cultivating proprietorship should be looked into as a serious obstacle to the mobilization of the poor cultivator for development.

V. INSEPARABILITY OF THE FARM AND THE HOUSEHOLD ECONOMY

Family farm arrangements also contain complexities that may make technological change difficult. The umbilical attachment of a family to a specific plot of land from which the members of the family draw the major part of their own food and a negotiable surplus to use in exchange for their other needs continues to be the basis of rural livelihood throughout the developing world. In principle, the farm is expected to maintain the family and the family to supply the labour needs of the farm. Unlike the capitalist farm, where labour is hired as and when required, the family farm must be organized around the imperative of feeding the family, which is its *raison d'être*, throughout the year. The productive round of the farm does not need the family labour throughout the year, while for certain types of crops (e.g. rice) additional labour may be required at certain seasons. On the other hand, food is not produced all the year round, and supplies must be stored from harvest to harvest if possible.

One result of this situation is that the peasant family, operating a farm too small to produce substantial reserves, necessarily incurs debts for consumption purposes, normally at high rates of interest. Adoption of the new technology involves an even deeper commitment. Already in debt for pre-harvest consumption and for occasional ritual obligations, the small cultivator faces the necessity of doubling or trebling his indebtedness if he is to change over to the new technology.

The locality studies show that such transformations do take place, but they are likely only when both agronomic and market factors are favourable and give assurance of success. Among these factors, an efficient inputs delivery system and the availability of low-interest loans play a very important part. The gravity of the problem is revealed by a glance at figures about size and distribution of landholdings in most of Asia and Latin America. These show that a large proportion of cultivators work farms that fall short of the livelihood threshold, and must therefore attempt to enter the labour market in order to maintain the health of their families. This condition is aggravated by a trend, caused by the pressures of large families, towards further fragmentation and diminution of the average size of holding.

At the same time, self-provisioning agriculture itself is weakened as a system of production and livelihood by the advance of monetization and the increased commercialization of the relations of production and exchange. More items of daily use have to be bought for cash as village crafts are replaced by manufactured goods. Sons require monetary compensation for work on their fathers' lands. Exchange labour is replaced by wage labour, and the spare resources that were available for emergencies and for the village destitute tend to fall under a stricter accountancy. And the increased need for cash, for purposes other than productive inputs, may be accompanied by a decline in farm size.

The studies throw up with great urgency the problem of the small cultivators, who constitute the larger part of the rural population of the Third World and whose numbers continue to increase. In addition, wherever the new technology increases the expectation of increased net returns, the price of land increases in such a way as to prevent the expansion of the smallholder, either by driving tenancies out of the market or by further adjusting exploitative conditions in such a way that entrepreneurial capacity is inhibited. In the product market, the small cultivators have to compete with large-scale producers whose costs can be lowered far below the point at which the small producer would fail, while in the factor market a variety of social dispositions put him in a situation of contractual inferiority *vis-à-vis* those who control access to inputs.

What this means is that the number of cultivators who may be described as small, in the sense that they own or have access to sufficient land to maintain family livelihood, may be stable or declining, but there is a larger and increasing number of cultivators who can only be described as marginal, since they have access to 'sub-livelihood' lands only, and must complement their own production with other income, from the sale of labour to larger cultivators, from trade, from crafts, from migrant labour, and so on.

In fact, the problem here is not that which is generally posed, namely the intractability of 'subsistence agriculture', but its decline to the point where it does not even provide subsistence. The movement out of agriculture implied in this process is a long-term trend that can be expected to take place, and which is a positive feature of development so long as alternative earning opportunities exist, but in most of the countries studied the alternative opportunities for productive occupation were so few that they were frequently insufficient to absorb the natural increase in urban population.

In these circumstances, the decline into marginality marks the inexorable destruction of the essential and customary means of livelihood of an ever-broadening stream of rural people dependent upon the labour market.

Village Polarization

It will be objected that our researches in new technology areas show that the labour requirement per acre in these areas tends to rise, except where favourable conditions for fuller mechanization exist. However, our study was unable to cover the course of events in the 80–90 per cent of areas where conditions do not favour the new technology, and it is here that marginalization will take place and where the marginalized cultivator is least likely to find adequate complementary means of livelihood. That is to say, unless governments can establish some system of continual sounding in order to register agrarian trends, the problem is capable of acquiring critical dimensions unperceived.

In fact, the process of polarization taking place in village life has a dynamic aspect that is more clearly revealed by two field examples. One of these, a study of the effects of various programmes of rural development in four villages in the state of Uttar Pradesh in India, shows how the class/caste hierarchy compartmentalizes relations in these villages and ensures the cornering by village élites not only of government facilities supposedly available on an equity basis, but also technical and other information diffused by radio and the printed word. The lower strata were found to have been excluded from places where public broadcasts were relayed and, being also illiterate, were still ignorant of most of the essential data content of welfare and development programmes several years after their introduction, including that related to the new technology and the facilities available for obtaining inputs and credit.[5]

A second study (*Franke, 1972*) is about a rice-growing area of Central Java. The whole region is characterized by skilled husbandry and a high level of cropping intensity in excellent climatic conditions, but it suffers from excessive pressure on land and offers little by way of occupational alternatives to agriculture. As a result, there is excessive land poverty and landlessness and a tight, unremitting struggle for existence. An intensive study was made by Franke of the village of Lestari in the year 1972, when he found that of its 266 families, 168 had no arable land at all or simply garden plots, and the total area in paddy, the main food crop, was no more than 75 hectares (which, however, was cropped twice a year)

---

[5] This was accounted for quite simply by the obvious conflict of interest in relation to the availability, docility, and price of human labour. So long as an important section of the rural population was landless, the élite, whose extensive holdings required labour, could obtain it at a low price. But official schemes to distribute state lands to the landless or to improve the entrepreneurial potential of the poorest cultivators were seen as damaging to élite interests, and were successfully opposed by maintaining a close monopoly on incoming information, inputs, and welfare measures.

and it was roughly calculated that normal harvests could produce sufficient food for 150 of the 266 families.

As a result of conditions like these, in many parts of Central Java the institution of labour debt has become widespread. Variations in family income of the poorer majority, whether derived in kind from the family's farming activities or in cash from wages or gains on petty trading, created occasions when food for survival was lacking. If the needy family were that of a cultivator whose harvest was still many weeks off, he might raise a loan in cash or kind in return for a promise to sell the harvest or by recourse to a 'green sale', i.e. by selling his crop while it was still growing in the fields at a reduced price to the lender-buyer. Or he might receive his loan in return for the pledging of his future labour to the lender at reduced wages. The effect of this institution, therefore, was to lower future production costs for the cultivators who disposed of surpluses, and this made it possible for them to accumulate further production capital.

For the cultivator with lands that did not attain the livelihood threshold, some security against periodic hunger was provided, but at the cost of decreased income from labour sold and decreased freedom to choose how to dispose of his labour. It was found that 109 family heads had incurred labour-debt obligations, and these supplied cheap labour when required to 17 surplus-producing cultivators.

Although the government programme introducing the new technology made credit available to cultivators, none of those who had incurred labour-debt took advantage of it. It is easy to see that this group, characterized by lack of reserves and vulnerability to seasonal fluctuations of income, would be dubious about substantial new debt-involvements, even though the prospects of greater gains at harvest time had been demonstrated.

From his over-all knowledge of the order of social relations in the village, Franke considers that a move by one of the labour-debtors to use the new technology by participating in the institutional arrangements of the government would have amounted to 'an act of political resistance to the control exercised by the large capital-holders'. He considers that 'the entrepreneurial activity of the poor would be a threat to the entire structure of privilege and security built up not only for the poor themselves but also for the wealthy who receive most of the rewards'.

Talents-Effect

Perhaps the most important aspect of the rural situation illustrated by these two case studies is the fact that rich and poor do not simply coexist. The accumulation of land by the rich creates a demand for labour, which the poor are obliged to satisfy because of their land-poverty or landlessness; moreover, the entrepreneurial success of the rich is made possible by the hunger and importunity of the poor cultivator, who is obliged to surrender his bargaining freedom and even to pledge his future labour at a reduced price in order to sustain his family and meet current obligations. Excessive pressure on land for cultivation and a steadily growing population intensify and dramatize the operation of this principle.

However, the descriptions make it clear that the economic peculiarities of land and labour are not alone in making exploitation possible: the advantages of the buyer of labour and the handicaps of the seller also rest on local class systems involving privileges and the consequent acquiescence by the majority (voluntarily or involuntarily) in the use of sanctions such as physical punishments, differential access to information about legal, administrative and financial systems by means of literacy and learning, the cornering of political influence, and on other social factors.

The interplay of the social and the economic that tends to strengthen the rich and enrich the powerful, as well as to weaken the poor and impoverish the weak, is so ubiquitous a phenomenon and so fundamental in its operation that some simple phrase is required that goes beyond systematic economic concepts and is more widely human in its application. We are therefore adopting the phrase *talents-effect* to refer to situations in which this process of polarization is induced. At base we are using this phrase to connote the most elementary axiom of competitive behaviour, namely, that the more talents the player has at his disposal, the more he is able to pile up—using talents to mean counters in the universal game of seeking a livelihood. The greater the number and variety of counters, the better equipped is the player, while the holder of few counters has the greatest difficulty in retaining those few.

While other societies, including many of the rich industrial ones, have developed mechanisms to control the momentum of this process, Third World rural societies undergoing further penetration by national and world market forces and 'modern' institutions are critically vulnerable. Economic growth or 'development' by a capitalist penetration of the prime zones of production and incorporation in the international economy appear to carry with them a dynamic poverty-generating principle unless the process can be understood and the political will is present to inspire governments to counteract economic polarization.

VI. INEQUALITY AND THE DISTRIBUTIVE SERVICES

Experience with development programmes poses a crucial question about the possibility of equitable access to facilities in a highly stratified society or a society with wide social differences in it. It is easy enough to see that in an open market situation the talents–effect is fully operative, and control over resources facilitates the acquisition of further resources, further control, etc. However, customs and institutions in most societies stand in the way of the fulfilment *ad absurdum* of this trend, and development programmes provide an example of such institutions. We find that agricultural services provided by the State are nominally available to every cultivator—the implicit principle being one of fair shares—and are therefore supposed to offset the dynamics of acquisitiveness. However, world-wide observations gathered together in the present study seem to demonstrate that such open distributive services (indeed one might say this of public services in general) operate on the basis of equity only where inequality is limited.

Equity-based programmes are introduced by governments based on constitutions that pronounce citizens' equal rights, and planned by officials who may aspire to a society having a truly common citizenship. Yet the realities of these societies manifest excessive inequalities and differences of life-style, a prevailing mentality that continues to accept and act upon such inequalities, and an attitude of reverence to a life devoted to the accumulation and appropriation of private property. The problem facing development planners is that distributive services and open institutions inevitably founder on such rocky bottoms.

The inequalities of rural society that turn open services into organs of discrimination are of several types. The simplest type of inequality is that based on the skewed distribution of landed property, which is directly concerned with income. To this form of economic inequality is added the fact that possession of land also makes possible access to capital.

A second type of inequality shown by our studies to be a critical obstacle to change and development is the dependence of the property-less on those who have property. This condition has self-perpetuating features, since the indebted ones see in dependence on their creditors some prospect of averting destitution, while the creditors (as individuals or as a class) find in the debtors a source of cheap and docile labour.

A third type of inequality is to be found where the population is differentiated along ethnic, religious, linguistic/cultural, caste, and class lines, and these sectors are considered to have differential rights, involving the domination of one by the other.

Very few of the situations studied were free from gross inequalities of one or more of the types given above, yet where such relative freedom was found, the services seemed also to operate more efficiently. This fact is commented on by Hameed (1977) in the Sri Lanka country study. Four localities were studied, one of which was a recently formed settlement (Minipe) in which cultivators had received equal endowments of land, and in which a common interest in the efficiency of the services seemed to outweigh individual temptations to subvert them for private advantage.

In the other localities studied, no serious caste divisions kept a sector of the population in handicap, yet it was observed that power and community decision-making had become somewhat withdrawn from cultivators and lodged in the hands of village élites consisting of officials, traders, and other groups who had been able to obtain control of substantial amounts of land, which they rented out.

Thus a relative inequality in land distribution, a system of tenancy that separated those cultivators unfortunate enough to be tenants from most of the benefits of their skill and labour, and a concentration of power in the village élite, all tended towards the abuse of the existing measures and services provided for agricultural development.

It is in the light of the patterns of inequality and dependence that the 'progressive farmers' approach needs to be appraised. The planners of the High-Yielding Varieties Programme in India were particularly interested in obtaining early successes with the new seeds in order to place on view a veritable jump in yields as a reward for using the new technology. The experience of the extension services during earlier activities and campaigns was that the advances they could promise were too small and too slow to galvanize the cultivator into changing his methods, increasing his investment and his work-load. They were also too insignificant to transform the *rentier* landlord into an active entrepreneur. Consequently a strategy was developed around the idea of the 'progressive farmer', usually but not necessarily one who owned a medium or large farm, was better educated, had already adopted some of the elements of the new technology, and had sufficient resources and financial backing to accept the risks involved in experimentation.

In fact, it was the resources and connections of these cultivators that led to their selection, and while they were progressive enough to accept the invitation, the implication that they were necessarily more progressive in outlook than small cultivators with poor connections must be resisted in most of Asia.[6] The strategy seems to have given the richer farmers a head start and, at the same time, made it more difficult to convince the small farmers that they too could

[6] The use of the term 'progressive' seems to have more meaning in Africa, where numerous observers have drawn attention to a profound change of outlook that a certain minority undergoes as a result of experience outside the tribal society, and which implies a rapid and enthusiastic embracing of new elements of technology, coupled with a rejection of communal authority, beliefs, and attitudes. (See, for example, *Weintraub, 1973* and *Feldman and Lawrence, 1975*.)

manage the new technology. It also accustomed the extension worker to operate through the rich farmer, whose more substantial livelihood made all things easier.

Concern for a broader rural development urges the consideration of a strategy focusing on the poor but aspiring farmer as initial adopter, since his success can have much greater influence on the majority than that of the rich farmer.

Jacoby (1973) feels that while very good results can be achieved through extension services in pilot projects and on a small scale, the cost of finding, training, and paying staff of the necessary high level is prohibitive when applied to a whole country for working with traditional smallholders. The Provisional Indicative World Plan (*FAO, 1970*) proposes a figure of one field worker to 500 farm families with a supervisor for every five field officers, and a smaller number of specialists. Other experts have suggested 200-250 farm families per field worker. Jacoby draws the conclusion from these calculations that the economic limits on efficient extension for small farmers are a major obstacle to agricultural development, indeed sufficient to warrant the adoption of collective or state forms of production unit, into which the extension function would be built.

Bearing in mind the requirements of the new technology in regard to the extension function, new approaches may be tried. As we indicated above, agricultural production now requires two kinds of technical knowledge—that of the field and its physical ambience, and that of the laboratory. The extensionist cannot be an expert in the laboratory, where a full response to a technical problem requires the work of several disciplines. He can, however be an expert in the field, especially if. his education has consisted partly of a struggle to produce livelihood from the field. The possibility of training a corps of field extension workers recruited from the peasantry (those who have been reduced to minifarms not requiring much attention) and whose identity with the peasantry continues, might be explored, since it has interesting implications for the mobilization of poor cultivators and also for the economics of extension work.

Regular Irregularities

The whole relationship between a government policy or a programme for agrarian development and the cultivator is managed and mediated by a staff whose interests lie in pleasing their employer—the government. It follows that the manner in which the relevant roles are performed is crucial to the successful transmission of the programme's change-inducing content.

While it is obvious that the quality of personnel running a programme is fundamental for success, it is a difficult and sensitive subject, requiring intimate and prolonged observation of a particular point of interaction between programme and peasant.

Mencher (1970) reports on such observations in India and what we learn from her is that the most vital link in the chain, the village-level worker, cannot expect progressive rewards in his career for faithfully performing his functions, and is therefore driven to seek the most advantageous personal position he can by means of a distortion of his professional work. With regard to the superior category of Agricultural Extension Officer, we learn that he is not so desperately shut off from promotion, but that his conduct is oriented by his solicitude for his own prestige (relevant to his promotion) and the importance of satisfying his superiors. This second concern expresses a profound dysfunction of the system since it militates against the transmittal of information about agricultural practice back to the centre of the organization, or to the laboratory.

The same kind of analysis, but carried further, was done by Sylvia Hale (1975), who made a careful and suitably quantified study of the performance of development programmes in four villages of what was considered to be a progressive district in Uttar Pradesh, India.

Situations of the kind revealed by Hale have of course been discussed, noted, and commented upon before but usually the observed pattern has been taken to be an 'irregularity'. The value of Hale's tough-minded interpretation of the state of affairs in her villages is that, given certain features of the social setting, such behaviour patterns are regularities, and predictable. Take, for instance, what we might call the 'compacted nexus', that is to say, the self-rewarding arrangements made by the representative of the government agency with the leader in the village. Both individuals have a nexus function to fulfil for those on whose behalf they act, but they compact to subvert their legitimate functions for their own profit.

Writing about the village he studied in Sri Lanka, Selvanayagam (in *Hameed, 1977*) says: 'The suspicion of younger members regarding the integrity of village elders and leaders also seems to be justified. Since most village leaders are land-holders and traders, they naturally have a greater influence in village matters. . . . Under the present set-up it is impossible for any welfare measure to seep down to the larger community; whatever small benefit that is intended for the community is quickly seized upon by this small coterie of men.'

While the government representatives are not here directly involved, elsewhere in Selvanayagam's report we find evidence of the compacted nexus in relation to seed distribution: 'Certified seed varieties are not always adequately available to cultivators. A few influential landowners manage to take the available seed paddy. Sometimes the seed paddy is adulterated. It was alleged that the local Agricultural Instructor (from a neighbouring village) used to favour "his" men, especially the rich landowners from his village.'

Luisa Pare's account of the same phenomenon (in *Pearse, 1975*) is based on observations made in several dozen credit societies in Mexico and the systematic self-benefiting arrangements made by the leaders of the credit societies and the official credit bank.

Danda and Danda (1971), in their study of Basudha, show a still unincorporated community in which one government official continues to be regarded as an outsider, whilst another, apparently as a result of an act of identification with the community members, comes to be accepted.

It is in a different situation again, and under different conditions, that the villager is expected simply to pay the government official for his services in the Swamp–Rice Scheme in Sierra Leone, and is described as follows here:

Apart from legitimate if excessive delays in distributing the scheme 'bonus' to participants, there is evidence that there are other, 'illegitimate', reasons why participants are not receiving their subsidy. An unbalanced relationship between the Agricultural Instructor and the farmer made it relatively easy for the former to gain advantage from the situation to the detriment of the latter. (*Weintraub, 1973.*)

Assessments of extension and community development programmes are inclined to refer to these distortions with some delicacy and to prescribe improved training and more careful selection. It must be faced, however, that the universality of these kinds of distortion reflects the fact that conduct cannot be regulated by moral norms assumed in planning field services, and individuals who are recruited to serve at different levels in government have many competing loyalties, which they consider as legitimate as that which they owe to the service.

In a profoundly unequal society in which the spirit of *laissez-faire* self-enrichment has been let loose, government office offers opportunities for individual 'development' that are refused only by exceptionally motivated personnel. And such motivation is most likely to be derived from political, ethnic, or religious solidarities.

In attempting to understand the forces that generate both wealth and poverty, it must be recognized that the privileges of office may be on the same footing as the possession of land, a strong social position, creditworthiness, a knowledge of the new technology or of the workings of bureaucracy; all may serve indiscriminately as 'talents' (coin) serving the 'development' of the individual.

VII. CRITIQUE OF THE GREEN REVOLUTION STRATEGY

What emerges from the evidence is inevitably a critique of Green Revolution strategy and not a rejection of the technology itself, the application of which can be widely beneficial under the appropriate conditions. The critique is here recapitulated, and in Chapter XIII a number of crucial issues are held up for discussion in the search for strategies that may fit individual country circumstances and conjunctures and also offer more humanly acceptable development paths.

It has been asserted that the package approach is frequently discriminatory since it calls on the cultivator to amend too many distinct aspects of his technology all at once, and to attempt a radical leap forward in which there is discontinuity between the existing and the new. Following visits to Thailand and the Philippines, Ishikawa has commented on the absence of a distinct phase of varietal comparison and improvements and pure-line selection among native varieties such as had taken place in the technological development of rice-growing in his own country (*Shigeru Ishikawa, 1970*, p. 6).

He insists that, in many cases, greater progress in peasant husbandry could be secured by initiating improvements at the point reached by the existing state of technology and developing the more scientific use of traditional inputs, or effecting a quality improvement of existing irrigation by the construction of terminal water distribution and drainage systems. This technological leap is the first hurdle at which the common cultivator, lacking the advantages of the élite or progressive farmer, is likely to stumble.

In contrast to these views, the Asian Development Bank's Agricultural Survey of 1968, echoing Green Revolution strategy, explicitly rejects the belief 'that there are development strategies that can provide enduring production growth by a judicious mixing of some aspects of modern science with the so-called realities of traditional belief and methods'.

Handicap of Size

The doubling and trebling of the cash cost of cultivating a hectare of cereal using the HYV package sets up a second discriminatory obstacle to pass: the cultivator must use his savings or borrow on an unaccustomed scale. But in fact only the well-off cultivators hold savings in normal years—the common cultivator is more likely to be in debt already as a result of his borrowings to maintain his family in the lean season or to meet the expenses of essential life-cycle ceremonial. Should he be convinced that agronomic and business success in the new technology will not elude him and that his gains will enable him to meet the additional costs of production, then he may borrow in order to pay for seeds, chemicals, and wages. Should he be fortunate enough to have fair access to institutional credit arrangements, there is a good prospect that his enterprise will be duly rewarded.

The majority of small cultivators, however, are likely to have decided already at an earlier stage that their life situation could not provide the necessary conditions for successful entrepreneurship in view of the known handicaps of poverty and the unknown hazards of the technology itself and of the external dependence to obtain supplies.

The discriminatory character of the package strategy and of the obligatory leap into capital-intensive commercial farming is aggravated by the selective 'progressive farmer' field tactic and the concentration of capital and technical services in already favoured areas. The result is self-fuelling pressure towards polarization magnified by a variety of political, social, and economic factors pushing larger cultivators towards a qualitatively more profitable agriculture and greater competitive strength in the market, accompanied by increased political power.

A misleading 'scale neutrality' was claimed for the new technology on the basis of the divisibility of seeds and chemicals, its main components. In fact, the socio-economic magnitude of the cultivator is of the utmost importance for his economic success, where he must compete with well-capitalized larger farmers.

Much of the outstanding success of the technology has been built around the control of water supplies through tube-well ownership, while the benefits of multi-cropping require tractor power to secure rapid harvesting and land preparation. Economically, smallness means absence of reserves with which to confront risk, and below a certain level it imposes the necessity of finding other economic activity to maintain the family throughout the year. Just occasionally, off-farm earnings are available at the right time as farm operating capital, but usually these are so exiguous that they are immediately absorbed for the purchase of food.

Finally, there is the probability that smallness is accompanied by powerlessness and dependence (through extortionate tenancy and debt) in a manner that interferes with effective entrepreneurship and bargaining power in the market place whatever the technology.

The net result, therefore, is that whatever may be the formal scale neutrality of chemicals and seeds, the great majority of cultivators are handicapped by their size in competing with cultivators who have ample access to land and credit. Advantages and handicaps are complementary to one another and polarization becomes cumulative—the talents effect is active.

X COPING WITH THE TALENTS-EFFECT

By way of illustration, various approaches to the problems posed by the talents effect (from India, Mexico, Malaysia, and China) are examined, and the essentially different character of the problem and the capacity to handle it are noted.

I. PROGRAMMES FOR 'WEAKER SECTORS' IN INDIA

In the case of India, special schemes have been set up for cutivators with a sufficient endowment of land to put the livelihood threshold within reach, provided they can improve the productivity of the land. Other schemes have been promoted for marginal farmers, for rural people affected by exceptional drought, and for landless people in areas of endemic unemployment.

The Small Farmers' Development Agency (SFDA) has elaborated a number of procedures for providing capitalization credit and other assistance to those small farmers who are potential surplus producers once they adopt improved techniques, input support, and irrigation. In some districts the agency has been able to take over a number of functions that otherwise would be carried out by the landlord: underwriting risks and providing certain entrepreneurial services, such as arranging for tractors where these are particularly needed for HYV rice farming. (*Parthasarathy, 1973*, p. 50.) At the time of the conference referred to below there were some forty-six SFDA projects, each of which was to be addressed to some 30,000 cultivators with farms of one to three hectares. (By February 1973 the SFDA had subsidized loans worth Rs. 8,884,000.)

A second agency (MFLL) was expected to elaborate projects for marginal farmers (those holding less than one hectare) and landless labourers. For the former, the projects were mainly to complement their deficient grain economies by the financing of horticulture, stock and poultry-raising, and dairying. Landless labourers were to be aided specially by the planning of rural works programmes that would contribute to the maximum exploitation of the agricultural potential in the area. By August 1972 the MFLL had provided Rs. 1.02 million worth of credit.

These programmes undoubtedly received an impulse from the introduction of the new technology once it became apparent that, for reasons already discussed, people in these categories could only in rather special circumstances make a real jump in productivity and profitability, and that in 1970 the proportion of the Indian rural population below the poverty line (i.e. the point at which the minimum calories necessary for healthy survival can be maintained) was estimated at between 38 and 50 per cent.

But there has been a considerable time lag in establishing the agencies and recruiting staff. In 1973 there was still only a skeleton field staff, so it took a long time to make the necessary identification of eligible farmers (*Kahlon, Sharma and Deb, 1974*). In addition, the state resources foreseen as supplementing central government funds have been inadequate and the programmes have thereby been adversely affected, according to V. S. Vyas (1973).

Two further programmes must be mentioned. One is the Drought-Prone Area Programme (DPAP) for rural people with low resource endowments and affected by droughts. Initially, the emphasis was on labour-intensive civil works of a permanent nature, but it has moved to strategies for integrated area development based on district level planning. Vyas wrote in the article mentioned above, however, that there was no evidence that this programme had fared better than the *ad hoc* famine relief programmes of the earlier years.

The Crash Scheme for Rural Employment (CSRE) is an emergency public works scheme for providing jobs for some 1,000 people in each district in the creation of durable assets: works in minor irrigation, soil conservation, afforestation, land reclamation, flood protection, road construction, etc. Nearly Rs. 320 million was spent on the scheme in 1971-2. But the States have found it difficult to keep the projects focused on long-term infrastructure, and they have tended to degenerate into a multiplicity of small, less significant projects, with attendant dangers of laxity of supervision and more wasteful use of money (*Vyas, 1973*).

Critical attention was given to these latter programmes in a conference held by the Indian Institute of Management in Ahmedabad in 1972. Not surprisingly, it has been difficult to get conservative local authorities, closely allied to the larger proprietors and employers of labour, to take up these schemes with energy. But the alternative, supported by many contributors to the conference, of setting up autonomous agencies to carry out the programmes, would tend to cause the existing farm administration bodies to withdraw still further from the majority sector, fostering an incipient dualism, a 'ghetto' sector, a 'bi-modalism' carried over into services.

Overshadowing the views expressed at the conference was a sense of uncertainty about the future. What was felt by observers to be lacking was over-all planning based on more realistic projections of the middle-future, i.e. the 1980s and early 1990s. What was foreseen as the destiny of the presently non-viable cultivators fifteen years from now? What were the expectations with regard to rural employment, in-migration, and urban livelihoods with the addition of a further 200 million people to the population? In the light of these questions, the programmes were felt to be patchy, and even where they lessened immediate misery, with the exception of the SFDA they did not succeed in contributing to the future earning capacities of the weaker sectors as they might if they could provide new productive lines and income streams securely grafted into the economy.

The observers have probably put their fingers on the central weakness of these programmes: they are an immediate response to critical localized manifestations of growing rural misery resulting from unbalanced growth, both as between rich and poor, and as between regions of differing resources, endowment, and population pressure. They do not confront the poverty-generating trends.

II. EXPERIMENTS WITH MINIFUNDIO CULTIVATORS
IN MEXICO

The Mexican study (*Hewitt de Alcántara, 1976*) describes a

case of extremely uneven development, with millionaire farmers emerging in the newly-irrigated semi-desert, as a result of the new technology and heavy government invest-ment, in contrast to a worsening situation for the individual cultivator and sharpening agrarian conflict in other rural areas of the country. The Puebla Plan is a response to this growing national problem.[1] It was also promoted by international foundations aware that many other countries (especially in Latin America) were experiencing the strains of a rapidly deteriorating peasantry, in the presence of developing entre-preneurial agriculture.

In the case of the State of Puebla, prior to the introduc-tion of the programme we are to discuss, a state of tension existed over agrarian questions. The number of landless labourers was disproportionately large. Fifty-three large estates (*latifundios*) had been permitted to develop, or survive, contrary to existing agrarian legislation. Between January 1967 and August 1968, twelve invasions of estate lands were reported in the press. While there was powerful opposition to the peasants' demands for a redistributive solution, funds were voted by the State government for providing credit and technical facilities. These moves by the State government coincided with a search by international agricultural experts for a site in which to work out a methodology for bringing about the transformation of subsistence agriculture under rain-fed conditions into a commercial type of agriculture, making use of improved technology.

The Puebla area corresponded to the needs of the inter-national group because of its numerous smallholder popula-tion, the possibilities of substantially increasing the production of the staple crop, maize, in rainfed conditions, and the existence of an outstanding market demand (in the city of Puebla) for surplus production of maize. (Puebla needed 425,680 tons per year and production reached no more than 239,440.)

The community of cultivators to whom the programme was addressed consisted of about 304,000 people, of whom

[1] Material in this section is largely based on a study provisionally titled 'Two Villages in the Puebla Plan' produced for UNRISD by Luisa Pare.

86 per cent depended directly on agriculture and the rest indirectly. Thirty-one per cent of the cultivators were small proprietors, 32 per cent held rights in *ejido* lands and 37 per cent were proprietors with *ejido* rights as well. The communities and many of the small properties resulted from the dissolution of large estates, and though properties had been equitably distributed, already pressures on land had made it difficult for sons to obtain land to work on, and there was a growing number of people whose share of land was so small that they regarded themselves as landless. Tenancies accounted for only 0.5 per cent. The average size of holdings was 2.47 hectares, only 1 per cent of holdings were more than 10 hectares, and about 50 per cent of holdings were less than 2 hectares.

The area was dominated by the capital city of the State, Puebla, with a population of nearly 600,000 people. Average family income amounted to US$ 505 and was derived from the following sources:

—the sale of crops	27%
—animal production	20%
—off-farm agricultural income	31%
—non-agricultural income	22%.

Seventy-seven per cent of the cultivators could read, and the average period of schooling was two to three years. The area was comparatively well supplied with roads and 63 per cent of the houses had electricity, though this played virtually no role in the productive unit.

The 1967 survey showed that only 39 per cent of the cultivators sold part of their maize crop and most of these did so under duress, i.e. in order to pay off debts or to meet doctors' bills or cope with other family emergencies. Maize was also accepted in the local shops in return for consumer goods.

The 'package' originally consisted of a variety of hybrid maize, appropriate fertilizer doses, and the necessary cultural practices, but it was found that the local maize variety already in use performed as well as the hybrid. The package, however, stipulated a higher plant density, with two applications of fertilizer in given amounts and at given times. The timing of the planting was left to the cultivator since, in many localities, cultivators are able to sow before the first rains by virtue of their knowledge of the moisture-retentive quality of the local soils. The changes proposed were therefore relatively simple, but the amount of fertilizer recommended multiplied the cash costs of production several times, and therefore required very important changes in economic habits. For most people these changes could not be made without obtaining credit.

The Plan originally set iself dramatic quantitative goals, one of which was the doubling of the average maize yield within five years. It was also hoped to raise the area 'technified' to 90,000 hectares by 1975, and to 30,000 by 1970. This latter figure was scaled down to 20,000 and performance attained only 12,500. However, by 1972, 17,581 hectares had been covered.

The figures for the increase in the number of participants and the area 'technified' do not fully reveal the influence of the Plan. Fertilizer was sold by a number of public and private agencies, so that it was not possible to check increases in sale and use of fertilizer by non-participants. However, Winkelmann (1972) was able to compare figures from the 1967 and 1971 CIMMYT Surveys and has given us a table showing a significant general increase in the use of fertilizer in the area (Table 15). It will also be noticed from this table that the difference between yields of participants and non-participants had been declining.

Since Plan Puebla was an experimental or pilot project designed to yield valid conclusions about the most effective methods of promoting the introduction of new technology, it is worth taking a critical look at the methods finally adopted before asking why results have fallen short of expectations.

As regards the 'package', it will be seen elsewhere that it was the subject of continual experiment and dialogue. The first step in promotion consisted of discussions with the regional authorities, who then brought the new technology and the facilities offered to the attention of cultivators. Next, demonstrations of the results of the experiments were given, to which were invited the authorities and 'key cultivators'. Of the latter, 103 were chosen to set up demonstration plots

TABLE 15

Frequency Distribution of Average Amount of Nitrogen
per Hectare on all Non-Irrigated Maize for a Sample[2]
of Farmers from the Plan Peubla Area

Average N/H	1967	1971
No nitrogen	38.2	28.1
1-19.99	14.5	7.8
20-39.99	19.4	15.1
40-59.99	18.3	13.5
60-79.99	8.1	12.0
80-99.99	1.1	8.9
100-119.99	0.0	7.8
Over 120	0.5	6.8
Sample Average (N/ha.)	29.4	57.1
Sample Standard Deviation	1,207.4	1,515.5

[2] Some farmers—those with no non-irrigated maize and those for whom it is impossible to separate fertilizer used on irrigated from that on non-irrigated maize —were eliminated from the random samples representing all of the area's farmers. There is reason to believe that this process has introduced a positive bias in the difference between the frequency distributions.

Source: Frequency distributions and calculations by D. Winkelmann (1972) from data taken in Plan Puebla surveys of 1967 and 1971.

(LARs: *lotes de alto rendimiento*), and provided with credit, technical advice, and insurance.

At the end of the season, each LAR was put on view for the neighbours with the participation of the cultivator-owner of the plot, who himself issued invitations to a demonstration meeting. General invitations were also given to attend the demonstrations by the Plan and the local authorities by means of public address systems, circulars, etc. At demonstrations a film about the new technology was shown, and for the purpose of obtaining credit for the purchase of inputs, groups were formed. A secondary approach was made by ensuring when possible that the fertilizer dealers were in possession of the 'package'.

By 1975, about one-sixth of the farmers had become participants in the Plan Puebla, and perhaps the same number again improved their technology in the direction of the 'package'. Why did these latter prefer not to join? And why are two-thirds of the cultivators apparently untouched?

An attempt is made to answer these questions.

Objects, Not Subjects

The ideal model of the Plan Puebla is given out as consisting of three components, namely the technical officers, the institutions and the cultivators. According to the considered view of the Global Two researcher, Luisa Pare, an anthropologist, the cultivators felt they were objects rather than subjects of the Plan, and indeed familiarity with the way the 'programme' was carried out showed that in most respects they were treated as objects. They had little voice in the planning of campaigns, in field trials and in evaluation of the results of 'packages'. Thus, the cultivators considered themselves as passive recipients of a technology developed outside the realm of their experience.

The participation of the cultivators in field trials of particular packages was limited to the loan of some land by the cultivators. In one village our investigator formed and sustained for several months a discussion group of cultivators, who went into many of the local problems. For this group, the experimental plot in the village was used for an experiment in hybrid maize by the Plan Puebla people, but no one knew what the results had been. The same ignorance of the results of the field research was manifest in the other village studied.

One technical officer explained the situation by saying that 'the researcher first selects a piece of land that answers to certain physical requirements and then finds out who its owner is so that he can borrow it for the experiment. Sometimes even the owner of the land does not know what we are doing there; it is our intention to explain it to him later.'

Similarly, the possibility of a more genuine participation was overlooked in the field of activity known as 'evaluation'. Instead of examining the results of the use of a certain packet of practices with the cultivators themselves and attempting to arrive at satisfactory explanations for success and failure, the evaluation was a highly 'scientific' one whose results were fed into the research station, where they were interpreted and used to modify the package.

Failure to enlist the skill, experience, and knowledge of the cultivators in the systematic thinking and planning to improve agricultural practice seems to have contributed to the permanent separation and distance that now marks the relation between Plan personnel and peasants and tends to solidify or freeze the attitudes of each to the other to the detriment of both.

This situation was revealed in many of the discussion group sessions held by the researchers. A large number of cultivators were experiencing crop losses from premature germination of the corn. But on account of the lack of communication between cultivators, and also between the cultivators and Plan Puebla technical officers, there was no putting of heads together, no systematic looking at the data in a way that might have helped the cultivators to find a valid explanation of the phenomenon.

In the discussion group, however, cultivators put forward their own explanations on the basis of individual experience. One explanation linked premature germination with the use of chemical fertilizer, another attributed it to too much fertilizer, and so on. Our researcher reports that members of the group manifest a 'natural tendency to experiment', which the Plan was unable to exploit or develop. So the peasants kept their distance and the technical officers tried to wrap up the technologies of the experimental station and sell them with the minimum of explanation to the occupants of the 'target area'.

Plan Puebla strategy, aiming at a general improvement of agriculture, took as its first target a single crop with which the cultivators were familiar, namely maize, and built its programmes around the promotion of the intensive use of fertilizer, for which most cultivators needed credit. It is suggested that the programme, by permanently coupling credit and the new technology, achieved an economical simplicity but also set up certain limitations, which were revealed in the two locality studies. In one of these it was noted that most of the small cultivators, holding one or two hectares of unirrigated land, were not in the Plan. This meant that they continued to plant maize by traditional methods when, with adequate financing, they could have probably tripled their yields.

One explanation given for their non-adoption was that as their small lots of land only required 40–60 days' labour per year, they depended on other occupations for their livelihood, relying on their plots for a store of subsistence grain for family use with the minimum risk taken and labour time expended. In particular, they chose not to join Plan Puebla, which seemed to have certain social and political overtones, and also implied certain obligations and activities connected with the obtaining of credit that clashed with their working hours.

In the light of land-tenure data for the area as a whole, it is clear that many of the small lots of land served as adjuncts to livelihoods based on wage-earnings or petty commerce. These part-timers could not be expected to view cultivation options as the full-timers did. However, having other sources of income, they might also have been able to fit into arrangements whereby the necessary fertilizer could be bought out of their wages, and without recourse to credit. The Plan, however, was not programmed to take care of the needs of people in this situation, which is one increasingly encountered in small peasant areas located near to poles of industrial or commercial growth (cf. the 'worker-peasant' phenomenon in contemporary Europe).

In the case of San Andres, one of the localities studied, the provision of institutional credit facilities for individual cultivators was not the most appropriate approach since the community enjoyed its own cash income, which could have been used for the purchase of fertilizer. The locality of San Andres was a land-owning community. Its extensive forest lands were rented out to a company manufacturing paper, for which the forest wood served as raw material, and the incoming rents could have been used for obtaining fertilizer in bulk.

The possibility of an *ejido* being responsible for guaranteeing payment of fertilizers, or actually purchasing them, raises another issue discussed at greater length elsewhere, the question of small fragmented holdings as an obstacle to efficient resource use. Mexico provides the interesting example of a country with a system of non-traditional collective landownership, and the requirements of modern technology have once again stimulated debate about whether or not Mexico should attempt to build collective productive units or perhaps collective services to family farms on the reality of collective land-ownership.

III. ACCESS TO CAPITAL FOR POORER CULTIVATORS: MALAYSIA

It has been asserted both explicitly and implicitly that the persistent generation of poverty is inexorably linked with the unequal distribution of productive assets, particularly of land, of capital, and of bargaining power as a function of political, economic, and social weight. In Mexico, the perception of this problem, brought to the fore by political instability, took the form of special programmes for *minifundio* areas. In West Malaysia, where most working producers of rice were also poor holders of small plots of land for which high rents had to be paid in many cases, the main thrust of the new technology promotion programmes was through the setting up of effective credit facilities.

The most highly developed and correctly weighted against the talents effects of these facilities seems to have been the system of credit centres latterly developed in Malaysia, of which an account follows.[3]

The Agricultural Bank (Bank Pertanian), established in 1969, assists paddy cultivators by lending to credit centres and by facilitating the services offered by the Federal Agricultural Marketing Authority. Because the Bank cannot put out sufficient personnel of its own at the local level it has chosen to use 'credit centres' as its agents, and these include farmers' associations, co-operatives, shopkeepers, rice millers or traders, and (since 1971) special bank officers for those

---

[3] Ingrid Palmer, who wrote this account, had an opportunity to look at the credit centres at close quarters while working for the project. It appeared in her draft report *The New Rice in Monsoon Asia*, UNRISD, Geneva, 1974. See general account of Malaysian credit in *Palmer, 1976*.

farmers who belong neither to farmers' associations nor to co-operatives.[4] Thus the credit scheme includes both public and private agents.

When a cultivator applies for credit to a farmers' association, the application is reviewed by the board of directors, and in the case of the Muda area, by the Muda Agricultural Development Authority (MADA) staff as well. Each association is divided into a number of units headed by a leader, who is a representative of the board of directors. The leader and deputy leader of the unit are responsible for distributing credit forms and collecting payment. Thus no farmer need travel more than three to four miles to obtain credit and extension services in the Muda area.

As a consequence of this delegation of banking responsibilities to local credit centres, certain contradictions have arisen, which are not resolved by Bank Pertanian urging the centres to lend much more.

The local credit centres are urged to accept all credit applicants but, if they insist on rejecting some, an explanation must be given to Bank Pertanian. The Bank has the right to order a farmers' association to accept an applicant if it does not approve the explanation. At the same time, once it grants an application the centre has to stand as guarantor for all borrowers. Since the farmers' associations constitute the vast majority of local credit centres, and because they are regarded as the main supporters of the weaker farmers, who may be rejected by the co-operative or shopkeeper, the quarrel is mainly between them and Bank Pertanian. At times, Bank officials hint verbally that the Bank would share liability with the farmers' association if something went wrong, but this has never been put in writing. A change of personnel or of spirit at the Bank is an important factor in the calculations of the farmers' associations.

Under the credit scheme, the farmer deals with coupons, a different kind for each type of input he chooses to purchase.

---

[4] This method of granting and distributing credit to cultivators is in marked contrast to that in Indonesia, where the State Bank insists on a direct (but costly) relationship with the cultivator. However, the absence of farmers' associations and the demise of the co-operatives in Indonesia does not make this scheme possible in that country.

After the coupons have been exchanged for goods, they are presented by input suppliers (including the local credit centres) for cashing. The farmer is, in theory, provided with sufficient credit to cover the full cost of enough fertilizer and insecticide for the recommended practices, but how much of the coupon book the farmer is advised to take or is determined to take is a matter of grave concern for those who are ultimately liable for his credit repayments.

One month is said to be required to process the individual credit application. The successful applicant has his coupon book printed with his name and identification number in Kuala Lumpur. The identification number is used to prevent duplication of requests since there is nothing prohibiting a cultivator from approaching more than one kind of local credit centre. Nor is there any means of stopping farmers from acquiring cash or loans for other purposes by selling coupons.

Bank Pertanian has set aside a large fund for paddy farmers that is not being utilized as fast as the Bank would like. On the other hand, farmers' associations understandably wish to lay down minimum conditions of credit-worthiness. The majority of loans made are unsecured but based on the board member's knowledge of the cultivator. If he has been with the association for some time, a loan is almost automatically renewed. More specifically, the association looks at the cultivator's yield record, his rent (which board members would probably know accurately) and the size of his farm. Rents are also verified by both tenant and landlord on the application form. The last is considered relevant in deciding whether the associations have had to rely on the judgement of fellow cultivators.

A screening committee is used to process applications for credit in the Muda farmers' associations, on which there is a representative from each unit of the association. Non-irrigated areas are eliminated first of all, because of the shortage of managerial personnel to cope with the applications. Then a limit of $300 per farm is set, which reduces most requests on larger farms automatically.[5]

---

[5] This limit has been opposed by Bank Pertanian and a University of Chicago consultant attached to the Bank.

The farmers' associations deliberately exclude credit for harvesting and transplanting. They also tend to limit the total sum granted so that very large farmers would get little on a per hectare basis. Bank Pertanian sets an upper limit of $261 per hectare. Without transplanting and harvesting credit, this is reduced to $113 per hectare. Thus the farmers' association maximum of $300 still represents a good-sized farm.

In the first season, 1970, $60,000 of credit was granted in Muda after two-thirds of the 3,000 credit applications had been eliminated. A shopkeeper acting as a local credit centre may not have the same criteria but would know of the farmer's over-all debt situation and his attitude to borrowing. In comparison, the co-operatives and private agents acting as local credit centres are generous to those whose applications they have accepted. The co-operatives, on the other hand, place more emphasis on land-ownership and the security it offers.

Interest rates are composed of a 3 per cent servicing charge to the local credit centre, plus a maximum of 9 per cent for secured loans and 3 per cent for unsecured loans. Out of this, the centre must pay 3 per cent per season in bank charges. Sometimes, the additional 3 per cent for unsecured loans is waived since the cost of enforcing and registering land title would be costly to the cultivator.

Against this, the co-operative bank lends to co-operatives at 4 per cent a *season* and the co-operatives add another 2 per cent a year. As the local credit centre, of course, the co-operative could use Bank Pertanian credit and charge the maximum of 6 per cent for secured loans. However, without offering extension services and low-priced inputs, it might not get many customers. The maximum charge of 3 or 6 per cent for farmers' association services includes hidden prices for these items. Therefore, when the co-operatives claim they can offer agricultural credit cheaper than the farmers' association, they fail to point out that the market is one of a 'differentiated good'.

Another cause for contention between the Bank and the farmers' associations is that in 1972 still only 10 per cent or less of credit was being used for chemicals in Muda, against 77 per cent of the credit for ploughing. The associations' reply is that, since farmers do not follow instructions on the application of chemicals, their use of them will not be optimal and so returns will not be as expected. For this reason, the extension officers hedge against advising great purchases of chemicals.

Although the farmers' associations (especially their extension workers) experience ambivalence over operating as credit centres because they feel protective to farmers and caution them against too rapid entry into an uncertain technological revolution, they are being pressured into 'selling the product'.

The extension officers of the farmers' associations are mindful of the meaning of the term 'credit absorptive capacity' and stress the importance of allowing each cultivator time to test his own farm's capacity before accepting a large amount of credit. Bank Pertanian argues that the faster the volume of credit grows, the more economies of scale there will be, the faster will be the farmers' associations' profits and the expansion of operations. But government agricultural officers attached to the farmers' associations are already greatly concerned that too much credit is being given for poor soils and that inability to repay loans will put the farmer in greater debt. Certainly the Bank, for all its ostensible generosity in granting loans, is very hard on farmers' excuses for inability to repay.

The policy of the farmers' associations in granting credit is to protect the cultivator from future indebtedness. They have expressed the idea that they do not believe in attempting to solve, in the short-term, problems that have been in existence a long time, and invite farmers to resign from the farmers' association if they are dissatisfied. They affirm that it is a better development policy to deny easy credit than to use police and courts to get money back. In Muda, there is now a problem of *over*-publicity of credit availability.

Whatever the argument for being cautious, the farmers' associations in Muda have an extremely low rate of unrecovered loans—3.5 per cent. This might be seen as the result of extreme caution, but it is worth bearing in mind, first, the liability of the association to repay in full and, second, that nearly 100 per cent of Muda association credits are unsecured. The associations have initiated a system whereby two farmers stand guarantor for a third. Although this is little more than a bluff, it has the social and psychological advantages of imposing a formal caution on cultivators for securing a loan, and it increases collective involvement.

Of course, failure to repay can be due to reasons other than lack of optimal input use and hard work. A cultivator claiming crop damage must report it to the Bank, which then checks with the farmers' association. Up to 100 per cent of the loan can be written off if it is a clear case, but this provision is not widely publicized. Over a large area (more than 4,000 hectares) the Bank demands that crop damage must be 'gazetted' before repayment is waived. Several causes of crop damage are defined. First, damage due to misapplication of fertilizers, etc.—no help. Second, damage due to pests—repayments are then stretched over a number of years. Third, damage due to drought, when the Chief Minister of the State has to declare the area an official disaster area—all debts are written off.

It has already been stated that institutions are seldom operationally transferable since their functioning is sustained by their links with a real and unique society. A number of aspects of the Malaysian institution can be thought about in this context.

Malaysia's export economy has been developed on a sufficient scale to allow a reasonable proportion of the budget to be devoted to rural needs and investment in the non-export agricultural sector. As noted previously the government is also committed politically to raising the level of the welfare of rural primary producers as part of the political pattern of managing an ethnically plural society. This gives a high priority to questions of rural livelihood on the part of a government noted in any case for the effectiveness of its public service.

It is also to be taken into account that, while the widespread existence of tenancy severely represses the net family income of tenants, at the level of the agricultural locality there is a considerable degree of equality amongst cultivators—and the percentage of holdings over 10 acres varies from 4 to 8 per cent in the different States. This means that domination by a land-monopolizing group is not an overriding feature of the local situation.

The farmers' associations described above have been able to acquire some importance, in the absence of excessive domination of local organs by large landowners, and these could effectively represent the interests of the small farmer. The conflict over acceptance of credit reveals the unusual phenomenon of the small cultivator being able to retain entrepreneurial freedom, and to take his time in commercializing the orientation of his farm according to his own livelihood criteria, choosing his commitments with prudence.

Even more noteworthy, perhaps, is the fact that the extension officers assigned to work with the farmers' associations seem to identify with the farmers' viewpoint. We have already expressed the view that the extension system should build its action and its propaganda around the long-term prospects of farmers' family livelihood rather than around production targets, or record-breaking deliveries of fertilizer per district.

IV. CONTROL OF TRENDS TOWARD INEQUALITY IN CHINESE AGRICULTURAL DEVELOPMENT

The path China has taken in respect of agricultural development makes no easy progress to obvious goals but is marked by experiments and changes of direction in pursuit of the solution of many and varied problems. Some of the great issues include the choice of investment priorities as between industry and agriculture; how far to rely on foreign aid and how far on rigorous self-help; the selection within the various traditional patterns of settlement of the correct basis for the collective unit of agricultural production; the desirable magnitude of family backyard economies to be allowed within the collective organization; problems of specialization of function and of the dangers of bureaucratization; the basis for income differentials; central direction as against local self-management; the problems of inequality of wealth between collective units and between different geographical regions of the country.

The Tenth Plenary Session of the Eighth Central Committee of the Chinese Communist Party of 1962 decided to undertake the technical transformation of agriculture by linking it with industry in a special relationship described as 'making agriculture the foundation and industry the leading factor' (*Stavis, 1975*, p. 95).

Prior to that policy crossroads, three periods can be distinguished in Chinese post-revolutionary agricultural development:

(i)   a period of rapid recovery from the disorganization of Civil War, from 1948-1952;

(ii) the period of the first Five Year Plan from 1953 to 1958 during which traditional resources—human labour, animal power, and natural organic manures—were developed to the fullest extent by the extension of double-cropping and a change of varieties of rice involving the substitution of *japonica* for *indica*;

(iii) the third phase, from 1958 up to 1962, when the Tenth Plenum took place, which belongs to the period of the attempted Great Leap Forward. During this period, in spite of the limits apparently reached heretofore, decisions were taken to intensify rice production still further by the kind of military mobilization of labour in large-scale units that characterized the communes of that period.

The attempt met with some initial production successes but the agricultural system was submitted to strains beyond its limits, as indeed was the social system that these economic arrangements imposed. Ishikawa (1972) asserts that the balance of the system was disrupted by attempts at increasing still further the intensity of cultivation at a time when intra-farm inputs were already scarce. Presumably he refers to natural sources of manure, to draught animals, and to human labour power, and he comments that the disruption was accelerated by a decrease in the number of both draught and meat animals. The same interpretation was suggested by Shillinglaw in a seminar held at the School of Oriental and African Studies, London, in 1969.

From both a technical-agronomic point of view and from a social one, the experiment was unsuccessful and the results threatened further progress. The use of the multi-village commune as a unit of production and accountacy was abandoned first in favour of the brigade (approximating to a large village) and then of the smaller and more intimately woven fabric of kin and neighbour families, the 'production team' with a membership similar to that of a neighbourhood group.

The post-1962 agrarian structure that has persisted until the present is based on a hierarchy of corporate bodies: the family; the production team, made up of persons brought close to one another by kinship and neighbourhood familiarity and working co-operatively to cultivate a farm; the brigade, in which the members of several production teams live together and enjoy the village amenities; the commune, which provides certain central services for the brigades and production teams; and beyond that the county, with its administrative functions in respect of a number of communes.

This system makes possible the insertion at the appropriate level of the necessary specialized installations and functions required by economic life and, of course, of educational and other social services. But each of these bodies is also capable of accumulating capital that can be invested in such a way as to benefit the whole collection of its constituents or, alternatively, to benefit one or more of them. Thus a mechanism exists whereby serious inequalities can be compensated.

While inequalities within the production team are likely to be transitory, responding to the cycle of family life and the age grouping of its members, inequalities between production teams and villages within communes—and differences between communes within the nation—are due to differences in the productivity of labour in different regions caused by the differential distribution of fertility and of resources, of levels of investment in infrastructure and transport, of the availability of power, etc.

Undoubtedly central, provincial, and county decisions about investment and location of industry are influenced by a concern to avoid sharp regional inequalities and the political problems they are likely to engender. Moreover, the most prominent structural features that seem to aggravate the inequality-creating mechanisms in the 'developing' market economies of Third World countries are lacking in contemporary China.

However, the talents–effect still remains a threat because of the momentum it can achieve. Quite significant differences in income and in wealth are to be seen at all levels.

Within the production team, the socialist principle 'to each according to his labour' is the underlying principle, and although there are certain differences between the value given to different types of work, the more important source of inequality between families is likely to come from significant differences in the proportion of wage earners to non-earning consumers in the family, and special cases where sickness and incapacitation reduce earnings. While minor differences may not arouse special concern, exceptional cases of poverty or destitution are looked after by the welfare fund, which is a regular cess on the net gains of the collective taken out each year.

The most important evidence of the absence of blatant inequality within localities is found in the apparent satisfaction of the peasantry with the government's policy. To quote a recent comment:

Vassily Leontiev, the economist, has commented that if the peasants have supported the revolution in China, it is because they have been given not merely long-term promises but more immediate and more palpable instalments of visible progress, and the promises have been kept. This policy has made it possible to prevent flight to the cities, the growth of urban shanty towns and beggary which are the common rule in the Third World . . . but the Chinese development policy is sharply contrasted with that of the other nations: it relies upon its own strength while the rest of the young nations are demanding reparations from their old colonial masters. (Alain Giraudo, *Le Monde*, 16 Dec. 1975, p. 37. Author's trans.)

Burki (1969), in his study of the People's Communes, refers to 'state assistance to less developed regions of the country', and to 'massive financial and technical assistance . . . to really poor production units'. He mentions also the diversification of the economies of poorer production units (presumably by external investment) and also by adopting a labour policy ensuring that recruitment for the non-agricultural labour force draws workers from poorer regions, thus raising resource availability per capita in the regions' production teams. A further measure that has been reported is a regulation obliging rich communes in areas suitable for the production of high quality commercial crops to continue to fulfil their own food needs by producing cereal staples rather than maximizing profits with the commercial crop and purchasing food crops from outside.

Wertheim (1973) suggests some change of accent since the Cultural Revolution. While equality as a goal of policy has not been abandoned, less importance is given to the reallocation of resources and to material aid from outside and more to raising political consciousness and stimulating renewed efforts at self-help and the achievement of self-sufficiency. More importance is also given to moral non-material motivations; numerous cases are reported of material gifts or the loan of experts and model workers from wealthier to poorer units.

Wertheim attributes the greatest importance to the fact that, in the world's most populous country, there is a constant struggle to cope with the problems of inequality and a threatening polarization: 'For what the Chinese are up against is not only old and outworn institutions and productive relationships; it is also human nature as taken for granted, as an immutable factor, by Western social scientists, economists and psychologists.'

Much can also be learnt from the Chinese practices in what may be called 'technological mobilization'. How are new cultivation practices, cropping cycles, agricultural machines, seeds, fertilizers, pesticides, and irrigation systems presented to the peasant members of production teams in such a way that they carefully examine the feasibility of the novelty and, once convinced, attempt to put it into practice? In other words, how do they manage the relation between scientific research, extension, and the organization of production? Without attempting any complete answer, some of the features of their working programmes are given here.

'The triple combination' is the name given to the mode of leadership in this activity, and it is a way of insisting on the necessity for a close and continuous leadership-collaboration between scientific workers, political cadres, and the masses. This means that scientists are expected to become directly involved in production. This has been realized by the establishment of 'demonstration fields', which, in addition to showing what can be achieved by the proper use of technology, are also large enough to make a significant contribution to over-all production. Some 'demonstration fields' have been reported to be as large as 1,000 hectares in area, and it was intended that they should cover 3 per cent of the cultivated area and bear crops with yields surpassing regional averages by 30–100 per cent. Management was put in the hands of agricultural institutions, whose staff were expected to devote half their time to them.

One of the purposes served by the establishment of demonstration fields was that of providing secure and adequate supplies of grain to cities. But they were also expected to provide a means of educating scientists in the conditions of large-scale production as well as in the social and economic conditions of rural China. At the same time, the political cadres working in them were expected to learn about modern scientific farming.

In 1965 it was reported that one-quarter (rather than one-half, as originally planned) of the agricultural scientists were to be found working in the demonstration fields. Even so, the criticism was made that fundamental agricultural research was slowed down as a result of the absorption of scientists' time in practical production. It was also suggested that scientists made poor farmers; this would seem, however, to be due to undue deference to their views by the peasants, whose more practical knowledge of local agronomic feasibilities should have had precedence over those of the scientists. (*Stavis, 1975*, pp. 156–72.)

In 1965 the experience appeared to have been positive and plans were made to extend the role of the demonstration fields. An observer who visited China in 1972 has reported that, in the province he visited, scientists spent one third of their time in the provincial institutes of scientific research, one third doing field research in the communes, and the remaining third touring and investigating techniques and problems in the communes.

At a different level, the extension service also operated through specific production units known as agro-technical extension stations, which were established all over the country from 1950 onwards. By 1957 they numbered 13,000, and with the Great Leap Forward they were transferred from county to commune leadership. The system tended to dissolve when the communes lost most of their powers in 1960-1, but were rebuilt after the Tenth Plenum in 1962 with five to ten such stations in every county, so that each one served a few communes. Stations were also established where there were special farms, and within State farms.

The purpose of the stations was to popularize useful scientific research and teach agricultural science to the peasants thus giving leadership to commune research organizations. They were also expected to solve specific production problems by bringing technical advice to the peasants in their production teams.

The stations were not responsible for original scientific research but were expected to test out suggestions made by the Chinese Academy of Agricultural Sciences as to their suitability to local conditions. The pedagogical slogan of the service was 'use facts to convince the masses'. The extension station was supposed to check novelties for at least three successive years before recommending them to ensure suitability in varying weather conditions.

Each extension station had two or three base points that were in fact usually production teams. It was crucial to get the right base point. One criterion was that it should be a team that volunteered, another that it should be given an income-guaranteeing contract, a third that it should be on land typical of the area served, and a fourth that it should be in association with the key point (*chung-tien*) of the Party Committee of the commune, to enable the extension station to benefit from better concrete leadership by the party.

The extension station was expected to provide one or two officers to be permanently associated with each base point—living and eating with the masses. At base points, experiments could be carried out and demonstration fields cultivated. In this way, the prescriptions handed down from the scientific centres had to stand the test of practice, with the representative of science living with the results of his recommendations.

Shared participation in the capitalization of agriculture at several levels appropriate to scale, reliance on local

resources and skills, and a close association between the
pragmatic problem-solving propensities of the peasants and
regional scientific research: these are three of the more
significant features of contemporary Chinese agricultural
development.

XI   CHOOSING THE RIGHT POLICY

It is argued that where inequalities exist already, the green revolutionists' strategy results in the persistence and generation of poverty for the majority of the people in rural areas.

Our conclusions are not intended to leave the reader in doubt about the historic importance of the developing genetic-chemical technology nor about its beneficent potential in the service of agriculture. However, there are two types of questions regarding future policy that call for discussion.

One of these has to do with the design and selection of the appropriate technology and the timing of its introduction into a country. It is obvious that a government must consider the relative importance of a land-saving capital-using industry-dependent technology against the claims of other technologies and other investment priorities. Mechanization, the intensification of a traditional extensive agriculture, or the improvement of small-scale irrigation, might all be more appropriate in that country or part of that country at a given moment.

Selection of the appropriate next step in agricultural technology is a subject of national importance in which governments should have a decisive voice, even if they are unable to control it fully. The decision depends on existing factor proportions, especially in respect of land and labour, the extent of infrastructural development, market demand for food crops, feasibility estimates of alternatives, and various other unique national and local considerations. It must of course be consistent with broad national development objectives.

The other type of question has to do with the institutional setting in which the technologies and their various components are introduced and the strategy by which an improved agriculture is incorporated in a dynamic social and economic process. It is about these questions that our discussion of policy issues ranges.

I. CRITERIA FOR POLICY EVALUATION

However, before pursuing these issues, it is necessary to establish criteria against which policies and their performances may be judged. Of the criteria set out below, the first rests upon the United Nations call for the 'elimination of hunger and malnutrition and the guarantee of the right to proper nutrition' given in the Declaration on Social Progress adopted unanimously at the Twenty-fourth Session of the General Assembly, 1969. Few governments would officially deny its primacy, though it receives little practical consideration when planning decisions have to be made, perhaps because decision-makers have little first-hand experience of the condition from which emancipation is sought.

The second is the clear consensus amongst the governments of the countries studied. They aimed at achieving food self-sufficiency (or rather, freedom from food-dependency) by means of increases in production and especially in land productivity to meet national needs and to keep up with anticipated population expansion. This policy was publicly justified both in terms of balance of payments considerations and of a desire to strengthen political independence. The third is an essential addition, made in order to avoid justifying well-fed autocracy, a viewpoint discussed more fully in *Strategy and Programme Proposals*, UNRISD (1977).

The three criteria are:

1) Freedom (for the people) from hunger and malnutrition (based on a minimum food basket);
2) freedom (for nations) from 'food-dependence'; that is, from the unwelcome necessity of importing food;
3) these policy aims to be pursued so as to allow an optimum degree of popular participation.

Are governments able to choose between policy alternatives? Are there no limits to such alternatives set by internal and external dependence and constraints? It is worth examining the realities of policy-making in order to be realistic about the relevance of our discussion.

Governments are dependent on the social groups that support them; they are not apart from society but part of it.

Instead of assuming that governments attempt to maximize some abstract 'national welfare' function but fail to do so because they lack the capability, or because they encounter 'obstacles', it is more fruitful for analysis to assume that they respond to the desires and interests of their different support-groups in the measure that these have real power to influence the making and implementation of political decisions. Thus power is exercised not only in determining goals but also in determining the means of achieving them.

Moreover, since the support-groups wielding effective power in any society have many contradictory interests, the policy aims and means adopted by governments are frequently contradictory and inconsistent to the extent that the government's power is shared among groups with opposing interests.

A second truism of policy making, following from the above, is that the overriding aim of all governments must be to maintain the power of the groups upon which they depend for their support. All policy goals and means are subordinated to this constraint.

It is not surprising, therefore, to find that even when laws embodying socially desirable legislation are passed, they are likely to be aborted in practice, or nullified by counterposed measures, if they are contrary to the interests of the support-groups.[1] A clear illustration is offered by the land reform

---

[1] It is sometimes not even possible for governments to make a straightforward choice of the most economically appropriate technology for a specific project. An official of the Harvard University Development Advisory Service has given an interesting account of the decision by the East Pakistan (now Bangladesh) Government in the late 1960s to buy equipment to build the irrigation tubewells essential to the country's agricultural development. On the basis of two earlier pilot projects—one using high-cost, high-power equipment and the other low-cost, labour-intensive equipment—calculations had been made of farmers' annual benefits and of internal rates of return to public and private investors. These led to the conclusions that low-cost wells were economically justified, medium-cost wells were less economically attractive but still justified, and that the high-cost wells were not economically justified. Indeed, the low-cost wells were the only type which had been proven to operate successfully in East Pakistan. But foreign aid donors agreed only to finance medium and high-cost wells, the equipment for which was more familiar to them and which presented fewer administrative problems, even though they were much less labour-intensive. The East Pakistan Government accepted a less than optimum solution, that of medium-cost tubewells, as the only practical means of acquiring the necessary technology. (*Thomas, 1973.*)

legislation adopted by many states of India in the form of laws granting property rights to tenants or limiting their rents and forbidding the holding of land above a certain ceiling by individual proprietors. The laws, enacted in response to policies that disregarded the interests of important sectors of the power base of many state governments, have been largely evaded by a variety of devices: tenancies have been disguised, tenants changed annually so that they cannot meet the conditions required for the transfer of ownership rights, landholdings above the state ceilings have been partitioned and distributed among members of the landowner's family, and so on.

A similar situation could be observed in the case of the agrarian reform in the Philippines decreed by the government after the *coup d'état* of 1972. As first announced, this reform made a resounding commitment to transfer ownership to the tenants of paddy and maize lands, predominantly holders of small plots of rented land of less than five hectares. However, it was discovered subsequently that the landowners involved were not a restricted group of land monopolists but included a wide middle stratum of small non-cultivating proprietors made up of the locally influential townsfolk: town and provincial people in business and professional life, and members of the administration—the very elements whose support was necessary to the new government. The government therefore amended the law in such a way as to relieve all holders of less than 25 hectares from the measures, and made the conditions of expropriation for the owners of more than 25 hectares extremely attractive. (*Kerkvliet, 1974.*)

II. ON HAVING A UNIFIED POLICY

Very few of the poorer and predominantly agricultural countries have been able to formulate and put into practice consistent and distinctive agricultural and food policies reflecting the broad national interest. Frequently enough, economic measures for different sections of the economy are inconsistent and even contradictory in relation to one another and to declared government development objectives. Policies are adopted without taking into account resource limitations in other sectors of the economy, while policies in other sectors frequently ignore the restraints on the expansion of agricultural production and the obstacles to the rapid reforming of agricultural institutions. Policies designed to aid one group of agricultural producers often hurt other groups. In most of the countries there was no really effective national agricultural development planning.

This incoherence in economic policies is perhaps greatest in countries with reformist styles of development where the balance among conflicting political forces is most precarious and the goals espoused by various interested groups are openly conflicting ones. Nevertheless, the same tendencies were observed in countries whose governments follow conservative modernization development styles or radical socialist ones. One need only recall the contradictions between the policy of sub-contracting the delivery of modern inputs for high-yielding rice growers to the Swiss-based transnational CIBA in Indonesia and the government's rural employment policies, or the conflict between the Cuban government's programme to produce 10 million tons of sugar in 1970 and its agricultural development policies aimed at domestic food self-provisioning.

Most Third World countries have the further problem of reconciling the powerfully represented interests of foreign and transnational corporations for production, trade and finance with nationally conceived policy aims. Perhaps, more than any other factor, it is this external dependence that underlies the inconsistencies of development policies. In fact, government power in some cases is exercised predominantly by an alliance of these foreign corporations with the national group whose superior urban livelihoods are most favoured by the corporations' activities.

III. CONTENTS OF THE FOOD BASKET

Just what should the family food basket contain? The answer to this question today seems to be simpler than it appeared some years ago. The 'protein gap' is no longer thought to be such an important and widespread cause of malnutrition. Ryan, Sheldrake, and Yadar (1974) refer to a report from the

Indian National Nutrition Monitoring Bureau on 1974 data from nine Indian states showing that the average protein content of diets in the lowest income groups was adequate, while in the two lowest income groups (comprising more than half of the population) the average calorie content was insufficient in all but one state. In another experiment, it was found that existing protein and lysine deficiencies revealed in children disappeared when their calorie deficiency was made good. Similarly, it appears that if the body's calorie intake is inadequate, any protein consumed is 'burnt' as an energy-giving food.

On the basis of less intensive research in West Africa, it would appear that in Guinea, Togo, and the Ivory Coast, average supplies of both protein and calories are adequate, while supplies of protein are adequate in Sierra Leone, Liberia, Ghana, Dahomey, Nigeria, and Cameroon, though supplies of calories are deficient in these countries by between 1 and 20 per cent. All nine countries were adequate in respect of lysine and amino-acids.

According to McLaren, writing in the *Lancet* (*Ryan et al., 1974*, p. 15), more and better protein is the answer to child malnutrition only where populations subsist on mainly starchy roots. These populations constitute only about 5 per cent of the world's malnourished, and reside primarily in the humid tropics.

The demonstrated predominance of cereals in nutrition also underlines the direct relationship between nutritional level and amount of crop land available to the family. It can also be asserted, on the basis of very extensive studies of 360 nutrition surveys from all over the world, that improved production in the self-provisioning section of the small farm is reflected in improved nutrition, while such improvement in production of the commercial crop has no such beneficial effects (*Schofield, 1974*).

As a result of recognition of the predominant importance of carbohydrates and of the fact that there is usually an inverse relation between yield and protein content, Ryan and colleagues, who work at the new International Crops Research Institute for the Semi-Arid Tropics, draw conclusions about research priorities for increasing food supplies. They put first on the list increased yields bracketed with greater stability of yields, via insect and disease resistance or tolerance. Next to these in importance, they propose increased attention to the digestible carbohydrate content (i.e. starch plus free sugars) of cereals, coupled with good cooking qualities. With regard to protein, in the case of sorghum this should be maintained at the 10 per cent level with the maintenance of the proper lencine/isolencine ration for the avoidance of pellagra.

In view of these facts, it is important that all countries should be fully aware of what proportion of their population depends directly on self-provisioning agriculture, either at family or village level, and take full cognizance of the fact that the health and strength of this portion of the population depends directly on the state of their agriculture. This important fact is liable to be left out of planning considerations.

IV. FOOD POLICY PLANNING

For those who wish to give a high ranking to an adequately fed and nourished population, a word of warning about the customary patterns of planning in market-oriented societies is required. The Green Revolution experience of the Philippines may be taken as an example.

Between the period 1961–5 and the year 1976, rice production in the Philippines rose from 3,957 to 6,439 million tons and yields from 1,257 to 1,808 tons per hectare. Rice planted using the new technology in irrigated land attained a yield level of 3.2 metric tons in the wet season of 1974 and 3.85 in the dry season of 1975. The improvement in yields and production was achieved by the use of a series of new varieties of seed, the application of increasing dosages of fertilizer, and the use of pesticides. There was a steady increase in irrigated land, and high yields in this land raised national yield averages.

Yields and over-all production suffered from the tungra blight and also from a typhoon in the early 1970s, but there was further improvement from 1973–6. In this latter period, a system of providing credit to tenant farmers without collateral was extended, reaching a reported 96 per cent of the cultivators, and by the latter date the amount of rice to complement national production was greatly reduced as national self-sufficiency in rice, the country's main food crop, was approached; great satisfaction in successful agricultural planning was expressed.

Yet the food situation of the majority of Filipinos showed no improvement and it was possible for responsible official sources to report that 60–70 per cent of all young people were undernourished. In fact, ten years after the beginning of a successful Green Revolution, the Philippines had one of the highest tuberculosis rates in the world. This must be taken as evidence of widespread nutritional deficiency.[2]

The danger is that increasing numbers of marginalized people, in the absence of cash income that can be transformed into purchasing power, are excluded from the planning process because they do not constitute 'effective demand'. Such is the duplicity of this type of planning that national economies can flourish while half of the population remains undernourished. The process of allocating resources, determining investments, and using other fiscal measures for guiding the economy has drifted so far from control of those forces that really determine human livelihood that new analytical tools are badly needed.

Price policy is one of the principal tools a government possesses to influence resource allocation, production, consumption, and, within relatively narrow limits, even income distribution. In most of the countries where studies were made, there was little evidence of a well-designed price policy that considered the whole range of agricultural product and input prices in relation to each other and to prices in the rest of the economy, with clear goals and criteria in mind. Instead there was a hodge-podge of price supports, price controls, and subsidies in response to pressures from different interest groups.

It should be recalled that the two principal functions of relative prices are to indicate the costs to the economy of scarce resources, thus helping to guide decisions affecting

[2] See *International Herald Tribune*, special supplement, 21 Feb. 1978; article by Bernard Wideman, with Leo P. Gonzaga, Mila Lahoz, Donna Reginsky, Bernie Ronquillo, Mamoru Tsuda, and Alice C. Valladolid.

resource allocation, and to reflect demand and supply in the market-place. While price policies are sometimes used to transfer income from one group to another, the limits of the price mechanism for this purpose are narrow. One man's price is another's cost, and vice versa, making the price system as a whole intricately interrelated. Manipulation of prices may easily produce indirect effects on cost structures, incomes, and resource allocation that are the opposite of those intended.

The difficult experiences of Chile, India, and Sri Lanka in attempting to enforce compulsory purchase of grains at prices below those generated by effective demand illustrate some of the problems that may arise; the large-scale contraband sales of Thailand's rice in Malaysia, where prices were much higher, is another example. In the Philippines the correlation of rice price increases with national elections was striking and apparently produced unexpected results on many producers' decisions. Where food grain prices have been held relatively low in relation to real livestock prices for producers, there have been large diversions of grain from human to animal consumption. 'Incentive prices' to producers that cannot be matched by additional consumer goods for which the money can be spent result in a black market and shortages. Relatively high 'unofficial' prices of one crop such as corn in relation to another such as sugar beet was accompanied in Chile by rapid substitution of the lower-priced crop for the higher-priced one.

Many governments have resorted to subsidies on farm inputs in order to stimulate the new technology. This was effective in Malaysia, the Philippines, Mexico, and many other places. However, it brings other problems. Since fertilizers and farm machinery must be purchased from industries that are usually located in the developed countries, when their consumption is stimulated balance of payments problems worsen. The use of locally available capacities and resources may be neglected and industrialization even retarded as a result.

As Keith Griffin (1972) emphasized in his UNRISD monograph, larger commercial producers usually can purchase their modern inputs at considerably lower real prices than can small peasant producers. This contributes to accelerated rural polarization as it enables those already better off to adopt the new technology rapidly and to gain control of more land and capital at the expense of the smaller tenants and land-owners.

International price policies are of course also extremely important in establishing the parameters within which governments can manipulate their own prices. The boom in commodity prices, especially of grains and sugar, and the depletion of world stocks, has created many incentives for entrepreneurs to produce for export. But to the extent that countries are dependent on food imports, they have faced severe difficulties.[3]

Conversations with scientists and planners associated with the promotion of the Green Revolution strategy reveal a view of society that underestimates the magnitude of the social impact of their initially technological revolution. It seems to be believed that the new profitability, supported by certain aids to entrepreneurship, can bring about technological transformations that will raise production and solve the problem of national food supplies. But a society at any given moment is a poised conjuncture of social forces, some in the ascendant, others emergent, and still others in decline; a new potential for profitability or a new source of power immediately become hotly contested objectives likely to be won by the strongest and the best armed.

When asked to contemplate the failure of technology-boosted production to improve the dietary of the common people, the green revolutionists are liable to blame the food distribution system, for which they consider themselves in no way responsible, but to which the government should turn its attention once it has promoted an adequate level of production. To the hungry, this sounds like the Red Queen's

---

[3] This whole complex subject is beyond the scope of the present study, but its effects on the spread and impact of the high-yielding varieties are very great indeed. In general, the trend towards rural polarization tends to accelerate even more in response to high world grain prices. Policies of the industrialized countries to protect their internal agricultural markets while encouraging their own exports creates additional difficulties for the poorer agrarian countries.

command to Alice: 'Jam yesterday, jam tomorrow, but never jam today!'[4]

What we have tried to show is that the character of a technology and the strategy used to promote its adoption operate very forcefully upon the whole social system, including those factors within it upon which its distributive function depends. The partial, production-oriented view of the green revolutionists inevitably strengthens the dominant groups within the agricultural sector wherever there are pronounced inequalities, which leads to the persistence and generation of poverty for the rural majorities.

As regards the sanguine view that governments can give every facility to the better-off entrepreneurs in the optimum agricultural areas and at the same time provide compensatory and welfare measures for the poor, this can be expected to happen only where fairy godmother governments rule countries with unlimited resources.

One of the most disturbing conclusions to emerge from the present study is the contradiction that appears to exist between the parameters within which development planners have to work and a humanely conceived view of development, which finds intolerable the continued hunger of so large a section of the population in face of the unforeseen potential of modern science. Just as the planning of production to meet 'effective demand' excludes from consideration the livelihood of all those who find themselves outbid for even essential foods or the wherewithal to produce them, so

---

[4] This dictum seems to underlie the views of Hopper, writing about 'Investment in Agriculture: the Essentials for Pay-off' in the Rockefeller Foundation publication, *Strategy for the Conquest of Hunger* (n.d.): 'The confusion of goals that has characterized purposive activity for agricultural development in the past cannot persist if hunger is to be overcome. National governments must clearly separate the goal of growth from the goals of social development and political participation. This is the most important pay-off essential. These goals are not necessarily incompatible but their joint pursuit in unitary action programmes *is* incompatible with the development of an effective strategy for abundance. To conquer hunger is a large task. To ensure social equity and opportunity is another large task. Each aim must be held separately and pursued by separate action. Where there are complementarities they should be exploited. But conflict in program content must be resolved at the political level with a full recognition that if the pursuit of production is made subordinate to other aims, the dismal record of the past will not be altered.'

economic model-making pre-empts discourse about national housekeeping and planning with a portfolio of concepts that find no place for human needs.

It is with awareness of this distortion of social science favouring the power holders and ignoring the weak that the UN Research Institute for Social Development has embarked upon the elaboration of the concept of a 'food system'.[5] The phrase denotes in the first place complexes or systems of human food-related behaviour, in which production is linked to consumption in all the different stages through which it passes as an object of transaction. The approach proposes the evaluation of food systems from the viewpoint of their capacity to ensure the adequate feeding of the whole of a population.

Such an evaluation can be conducted in terms of the efficient use of local and regional resources for national food supplies within the constraints of available energy and technology. It can be used to identify or reveal the obstacles to the various food-circuits on which all states depend and to reveal the existence of ineffectively used resources, and also of key-points in the system where R & D (technological research and development) is most required, whether in connection with production, storage, conservation, transport, distribution or kitchen practice.

The examination of food systems can also reveal problems of seasonal variations in supply levels and control of buffer supplies. On the basis of full multidisciplinary studies of food systems, the political means can be sought to frame and implement food policies. But the most significant feature of the approach is that it is to provide the research and information basis for an essentially humane criterion of development around which mass political support may be mobilized and economies set to work deploying science and technology for the health and strength of the many rather than the stock-piling of the few.

V. WHICH STRATEGY?

What contribution can this publication make to an improved

---

[5] See *Food Systems and Society* (project proposal), UNRISD, June 1978.

strategy and to decisions about appropriate productive technology? The last two decades have seen an overwhelming output of literature about development, supporting careers in the development industry, but one must ask at what point and with what effect have they connected up with history, with the histories of the so-called developing countries?

Neither the earlier development line promoting industrialization and input substitution nor the more recent adoption of agricultural development and promotion of the new technology has halted or slowed down the slippage towards pauperization and marginalization. An ever-increasing proportion of the rural population lose the connection with the land and its potential for feeding them, and fail to obtain any alternative food entitlements.

The planners believed that the problem of the poor could be adjourned until economic growth had been engineered. But the planners were largely irrelevant to the career of the nations, pushed along their stormy courses by the dominant élites representing minority interests.

Having declared a commitment to a policy that puts first the adequate feeding of the whole population, what contribution can be made towards its realization within the political context of each nation? Not certainly by Utopian proposals, nor by suggesting distributive measures that have been meaningful in particular national scenes.

Our conclusions, therefore, must be an offering to those whose thinking will sooner or later find political expression and will influence decisions from outside or from inside. They may chance to catch moments of institutional change, to enter into the basic thinking of future bureaucrats through their training, or become elements in political platforms, or provide utterance on behalf of strata condemned to silence, or be taken and used merely as a smoke-screen.

Their purpose is to strengthen those forces likely to press for non-élitist development. With this intention, we shall now return to the discussion of some of the implications of agricultural technologies themselves, and pre-conditions for their general adoption, followed by the underlining of a number of strategic principles or orientation favouring the stated policy criteria. In order to introduce greater realism, reference will

be made to the development experience of Japan, China, and
Taiwan, all of which have adopted peasant-based development
paths in agricultural and food production and have achieved
or are on the way to achieving high levels of land productivity by the implementation of science-based technology.

XII APPROPRIATE TECHNOLOGY

Rural livelihood is stressed as a leading criterion to decide whether or not a particular technology is appropriate; and land-and-water improvement is identified as the most vital single element in increasing food supplies.

There are three main directions in which investment in tropical food crops may go: towards land-and-water improvement, towards high-yielding varieties and fertilizer, and towards mechanization. The choice of a technology composed of elements of each of these dimensions is influenced by deliberately-made decisions by governments and their advisers, by cultivators, and by agribusiness promotion, though very complex political forces are also likely to be implicated.

In this chapter, a number of comments are made that are considered relevant to how these decisions can adversely or positively influence rural welfare.

I. LAND-AND-WATER IMPROVEMENT

Water is a fundamental element of technology, and the water that is truly valuable to the cultivator is that which is available in the right quantity and place, and at the right time.

The first attempts to introduce the new technology into a particular area often take place under the most favourable conditions, producing initial high yields followed by a decline in average yields as use of the new seeds is extended over land having inferior conditions of soil and irrigation.

Water is distributed by means of elaborate canal systems covering large regions, of open wells with primitive pumping arrangements, of tubewells, and rain catchment networks, by pumped underground water, and by means of combinations of these. But the efficacy of these arrangements is jeopardized by the irregularity and the fragmented character of the lands composing individual farms, by technical shortcomings, by inefficient and unco-ordinated management, and by lack of co-operation and co-ordination among users and suppliers. Canal seepage, local water-logging, and alkalinity are frequent.

In many rice areas, irrigation is carried out by the field-to-field method, without the use of any water courses at all. Frequently, distribution from the canal outlet to the farmers' fields is done through unlined ditches, allowing excessive wastage. Inefficient irrigation is not only a question of poor physical infrastructure: in most parts of the world, cultivators are excessively prodigal with water when they have access to it and ruthless towards neighbours in the struggle to obtain access. An efficient irrigation administration must be realistic about these cultural traits and work towards local co-operation for efficient management within a framework of firm and equitable control exercised by an authority.

In much of Asia it could be claimed that the type of investment that best contributes to raising the productivity of the land and to the social benefits of the area consists of labour-intensive works aimed at the improvement of existing systems of irrigation. Land improvement for better irrigation consists of the levelling and shaping of each field for uniform application of water and the avoidance of waste, the construction of lined and properly aligned canals, and the provision of drainage. This must be done in a comprehensive manner over fairly large geographical areas, so that the planning and execution of the major and medium drains may keep pace with the construction of field drains. Indeed, the most logical approach is that of the total restructuring of sections of old and poorly exploited irrigation systems, accompanied by the equitable rationalization of fragmented holdings and redrawing of property boundaries.

It has been conservatively estimated that if the 40 million hectares of irrigated lands in India could be made to produce three tons per hectare of foodgrains, the country's food needs could be met, leaving much of the remaining 100 million hectares free to be put to more rational and profitable use.[1]

---

[1] See 'A Charter for the Land' by B. B. Vohra, Chairman, Central Ground Water Board and Joint Secretary to the Government of India, in a paper circulated by the Ministry of Agriculture, New Delhi, September 1972.

This view is strongly supported by other distinguished experts (*Carruthers and Clark, 1980*) who see in the increased efficiency of existing irrgation systems and the exploitation of known groundwater resources the greatest potential source for increased levels of food supply in the immediate future.

Perhaps the most astonishing fact about water is that though its importance for organic growth equals that of soil, so very little is known about its behaviour in the fields. Even the new technology programmes tend to take water availability for granted, though no one doubts that a very large part of the failure to secure anticipated yields is due to confused and uncontrollable systems of irrigation. Even today there is surprisingly little knowledge of how water moves after being released from the canal, or what arrangement of the canals promotes efficient water use.

On the basis of Japanese experience with rice, Ishikawa (1967, 1970) takes the view that most of South and South-East Asia is not yet ready for the seed/fertilizer technology, with the exception of a restricted number of areas where the necessary investment has been made and where land-and-water development has been brought to a certain level of maturity. His position is that where yields remain below a threshold of 2.3 tonnes of paddy per hectare, investment in land-and-water development should be given priority. This should consist of efficient flood control, with the construction of irrigation and drainage canals. These preliminary measures of water control are, he considers, the true prerequisites for the introduction of the seed and fertilizer technology.

In Japan, extensive land-and-water improvement works were carried out during the Tokugawa period, when autocratic structures made possible the indispensable command over extensive drainage areas and a numerous labour force. Technological improvements involving the use of selected and improved varieties of seed, increased fertilizer use, and certain accompanying refinements of husbandry were promoted by the reforming Meiji state and taken up by peasant cultivators, both tenants and small proprietors.

In Taiwan, as described in an earlier chapter, the Japanese colonial power spent the first twenty years of the present century searching for a rice variety that would be acceptable for exporting to the Japanese consumer, pushing the extension of canal irrigation and improving husbandry. The colonial institutions were similar to the organs of government in Japan itself—and it was possible to make use of experience in rural administration and agricultural development gained in Japan.[2]

In the People's Republic of China, a very large investment of human labour and other resources has gone into the establishment of efficient irrigation, culminating in massive mobilization of labour during the Great Leap Forward, to be followed by planned concentration on mechanical control of irrigation canal systems by electricity, and their supplementing by means of extensive tubewell boring and the installation of low-lift pumps. The order adopted in raising grain yields operates by stages, with a first step in which land-and-water arrangements are improved, followed by the transformation of parts of these 'improved regions' into 'high and stable yield areas' in which water control is adequately mechanized. It is in these latter areas that the new technology becomes a priority.

Land-and-water improvement has been effected in the three countries in question by virtue of command over natural irrigation areas, and over large contingents of labour. Through historical circumstances, in the one case this was provided by harsh feudal relations of domination; in the second by colonial coercion and an accommodating traditional kinship structure; and in the third case by a revolutionary government able to mobilize the poor majorities by convincing them that collective action would improve their livelihoods.

The main obstacles to widespread land-and-water improvement are the high cost when these are financed by direct

---

[2] Use was also made of the *pao-chia* system, a customary grouping of families roughly in tens so that control of the conduct of members could be exercised and civic duties properly carried out. Instructions about new productive methods and the organization of work could be channelled through *pao-chia*, stiffened by the use of the rural police for extension work. *Pao-chia* also carried with it a cultural imperative of 'dutifulness' that made for fulfilment. There was coercion and occasional conflict, but the general trend of the colonial government's measures was beneficial to peasant families.

government investment, and the problems raised by the unequal and fragmented ownership of lands, and the complex structures of exploitative tenurial relations erected upon that base. The practical and economic approach to reform requires the mobilization of the labour of the rural population affected by the proposed improvement and the use of local materials. However, the incentive to such mobilization must be an entitlement to share in the use of the renewed and irrigated land, not a return to the former exploited status, and few governments can embark on so radical a policy.

Probably the most important development in irrigation during the last decades and that which has made the greatest contribution to the new technology is the use of ground water, made possible by the installation of tubewells worked by diesel or electric motors. Aside from problems of electric current fluctuations and fuel cost and scarcity, tubewell irrigation provides a controlled water supply.

The most spectacular successes of the HYVs of wheat and rice in the Punjab have been obtained with the use of tube-wells. More recently, in many rice-growing areas of monsoon Asia, where water is available only for a part of the year and conditions for HYVs are unpredictable or simply indifferent, tubewells are being used for irrigating dry season rice with excellent results. However, because of the size of its command area and the initial cost, the undivided ownership of tube-wells is restricted to larger farmers and the rapid spread has taken place without control or plan.

Mention should also be made of the convenience of community ownership of 'medium-lift pumps' for drawing water from existing open wells in dry areas where it can be found at 5-20 metres of depth. Worked by diesel engine, they are portable and can be taken from well to well as needed by cart, one such pump serving 10 to 20 wells.

Various approaches may be adopted in order to assure the widest possible spread of the benefits of government invest-ment in irrigation. On the one hand, credits for irrigation work may be restricted to areas of smallholdings, with the strict application of a low land-ceiling. Alternatively, owners of unirrigated lands could be granted irrigated land only up to the value of the unirrigated land previously held. Or again, all land endowed with optimum irrigation could be given out by the command area authority on rent and all former owners of land could be paid rents at pre-irrigation rates.

With the début of the International Crops Research Institute for the Semi-Arid Tropics in India, attention is being given at a high level, though on a very limited scale, to research in water use and management. One of the important problems under study in this Institute is that of the run-off of flood waters, a necessary prelude to serious investment in semi-arid lands. Low-cost water storage and a number of other aspects of water management are also being studied. But the subject is one to which cultivators themselves and specialists in a variety of areas may be expected to contribute.

It is perhaps through well-planned investment in minor irrigation and the improvement of existing irrigaton systems, making use of labour-intensive methods and coupled with measures to redistribute the new wealth added to the land by the investment, that solutions may be found towards increasing food supplies and improving rural livelihood, with the introduction of the seed-fertilizer technology in due course.

At the same time, research in water economy and technology is a matter of urgent concern. It is ironical to reflect that neglect of water research is probably due to the difficulty in establishing private ownership of water in order to commercialize it. But it would be tragic if we had to wait until it became a scarce resource for its conservation and management to receive the seriousness it deserves.

II. ACCESS TO FERTILIZERS

A series of wider objections to the heavy commitment of the new technology to manufactured plant nutrients and protective chemicals based on fossil fuels have arisen as the 'energy crisis' has proceeded. The rise of 500 per cent in world fertilizer consumption between 1948–50 and 1972–3, and the over-all trend towards high-energy-using industrialized agricultural systems raises very serious questions for the world community.

The 'energy crisis' has already resulted in greatly increased fertilizer prices and severe shortages occurred in many areas.

Many poor countries with foreign exchange difficulties have had to cut back on imports. While the potential availability of nitrogen is practically limitless (although the potential supplies of phosphate are not), the world's capacity to produce more and more energy at an exponential rate is questionable both from the standpoint of energy sources and from that of possibly irreparable environmental damage.

John and Carol Steinhart (1974) have estimated that, while in traditional food systems only about 0.5 calories of energy are required to produce one calorie of food, in the modern US food system one calorie of food implied the use of about nine calories of energy in 1973 as against only one calorie of energy in 1910.[3] The long-run implications of industrializing food production in Asia, Africa, and Latin America are staggering and should be studied carefully by the international organizations.

Leaving aside these long-run problems, the immediate effects of increasing demand for chemical inputs are serious enough. How can Indonesia, India, and Bangladesh obtain the necessary capital to build their own chemical industrial complexes or obtain the foreign exchange and credits needed to import chemical inputs in ever-increasing quantities? What will be the effects on their development plans of diverting more and more resources to this purpose? What are the implications for employment, income distribution, and institutions?

One of the ways in which a solution to the problem of the scarcity and expense of fertilizer is being handled in the poorer countries is by a search for alternative sources of plant nutrient that can keep manufactured fertilizer requirements to a minimum. According to recent observers talking to scientists in South India and Sri Lanka, a wide variety of such alternates are already the centre of discussions, tests, and research.[4]

---

[3] See also Pimentel et al. (1973), who draw attention to the advantages in terms of energy use and costs of a return to the use of animal manure in the United States. He and his colleagues assert that one milk cow or four hogs could in one year produce the nitrogen needs for one hectare of maize under single cropping.

[4] Among possible substitutes for the manufactured article were: green manure

III. MECHANIZATION

Our observations make it clear that each country requires a carefully considered policy towards mechanization in which full weight is given to rural livelihood as well as to sources of fuel and the manufacture of machinery. The issue is put this way rather than in terms of employment alone since a labour-saving device introduced into an entrepreneurial farm may increase the entrepreneur's profit and deprive rural workers of wages and consequently of livelihood, while the same device used by a working cultivator may increase his leisure (or free him for other productive tasks) without subtracting from his livelihood, but rather adding to it.

In some countries, barter and purchase arrangements for the importation of machinery have been made in the context of larger trade agreements, or as a result of persuasive salesmanship, without serious consideration being given to the socio-economic implications. In others it was found that fiscal and exchange rate policies effectively subsidized the importation of machinery, which was both labour-displacing and favourable to large rather than to small farmers. (*Griffin, 1972.*) This implies a form of control directly beneficial to a

---

(albecia leaves, etc.), cowdung, citronella ash, burnt paddy husk, nightsoil and urban compost, tank mud, other composts, seeding paddy fields with blue-green algae, seeding and harvesting blue-green algae in irrigation tanks, and the development of N-fixing pastures with seeding done in tank beds. Other approaches to the problem involved smaller amounts of manufactured fertilizer in combination with elements already available locally, or capable of being cultivated. These included labour-intensive fertilizer placement in the soil, carefully timed and split N applications to decrease N losses and provide N at highest response times in the life of the plant; research on relationship between nutrient uptake and season; the introduction of legumes into cropping patterns; closer spacing to reduce weed growth and loss of N, P, and K to weeds; closer spacing to increase N fixation in the rhizosphere of paddy plants; improved N-fixing micro-organisms for the paddy rhizosphere; combining biological N-fixation (blue-green algae and root micro-organism) with chemical N by applying the chemical N (which inhibits biological fixation) late in the life of the plant; breeding with an eye on the N response curve; preferring varieties in which the initial rise in calorie yield is high; and slow-release N fertilizer. Work is also being done on the improvement of customary rotations and the finding out of new ones. (The list comes from a paper presented by Robert Chambers (1974) at a seminar on problems of rice-growing areas of Tamil Nadu and Sri Lanka by the Centre of South Asian Studies. Barbara Harriss also contributed to the list.)

particular group but contrary to the general improvement of livelihoods.

A policy of controlling the importation and manufacture of machinery according to criteria of appropriateness for rural development requires also a complementary programme aimed at the encouragement of localized R & D for the improvement of installations and tools, with or without the use of power. Some work is being done to improve the efficiency of animal draught power for cultivating in India, and to encourage its replacement of hoe cultivation in parts of Africa. An interesting example of successful low-cost technology arising out of local inventiveness is the bamboo tubewell.[5]

Work of this kind has been pioneered in the international research institutes but the effort requires a popular dimension, and a link between small farms and market town workshops. An intense and broad-based movement is needed for refining and popularizing technologies appropriate for small units using family labour. However, it must be recognized that the exercise of the necessary controls implicit in the notion of 'selective mechanization' requires great tenacity of purpose on the part of governments, and examples of such controls successfully exercised are hard to find. Well-considered advice from United Nations agencies about the effects of mechanization in specific cases and about the possibilities of selective mechanization should strengthen the hand of governments especially concerned about rural employment. Advice should be obtainable from international agencies such as UNIDO, ILO and FAO, as well as from the international research centres. There are also a number of national and private agencies whose service might be enlisted.[6]

Further ideas for appropriate low-cost machinery and

[5] See P. S. Appu's article, 'The Bamboo Tubewell' in the *Economic and Political Weekly* (Vol. IX, no. 26, 29.6.1974). Invented by an Indian 15-acre cultivator, it can be made by the village blacksmith and requires a pump-set and a strainer made of bamboo strips and coir-string. Its cost is about Rs. 250 instead of Rs. 4–8,000 for an iron tubewell.

[6] The *OECD Observer* (No. 75, May/June 1975) published a list of the main intermediate technology centres, as follows: Intermediate Technology Development Group Ltd. (London); Brace Research Institute (Ste Anne de Bellevue, Canada); Appropriate Technology Cell (New Delhi); Planning Research and Action Division (Lucknow, India); Appropriate Technology Development Unit

other elements of technology are likely to emerge from interaction between engineers, scientists, and primary producers.
It is also worth considering the transfer of intermediate
technology from one poorer country where it has proved
successful to another. Chambers has, for instance, raised the
pertinent question why the bicycle trailer, so important a
means of transport in Asia, is not seen in Africa. Jequier,
author of the *OECD Observer* article referred to in the foot-
note, considers that the primary school is the place where
potential inventiveness might be fostered, by the study and
use of local technology for teaching more general scientific
and technological notions. He also points out the need for
credit to be made available to develop promising local ideas.

IV. INTERNATIONAL CENTRES FOR RESEARCH
AND TECHNOLOGY

Very few Third World countries are in a position to set up
national research establishments able to meet the main
scientific requirements and the special needs of the genetic-
chemical technology. As we have seen, a very important part
of the developmental research needed for the new technology
in wheat and rice was the product of the International Rice
Research Institute (IRRI) in the Philippines and the Centro
Internacional de Mejoramiento de Maíz y Trigo (CIMMYT) in
Mexico, although Japan, China, Sri Lanka, and other countries
have been mainly responsible for the development of their
own technologies. The activities of these centres can be
expected to continue to supply essential needs of non-
industrial tropical countries in respect of the multidisciplinary
basic research in those particular food crops for which IRRI
and CIMMYT have become famous. This implies working

(Varanasi, India); Industrial Development Division, Engineering Experiment
Station (Atlanta, United States); Volunteers in Technical Assistance (Mt. Rainier,
United States); Division of Microprojects (Eindhoven, Netherlands); Appropriate
Technology Centre (Islamabad, Pakistan); Technology Consultancy Centre,
University of Science and Technology (Kumasi, Ghana); Agricultural Engineer-
ing Department, the International Rice Research Institute (Manila, Philippines).

(Further information is available from the OECD Development Centre, 94
rue Chardon-Lagache, 75016 Paris. Also available from the Centre is a report on
a seminar entitled 'Low-Cost Technology: An Enquiry into Outstanding Policy
Issues'.)

with a vast range of genetic material, and sometimes the making and testing of hundreds of thousands of crosses in a single year; the building up of collections from all over the world of germ-plasms for the crops for which they are responsible, and clearing-house functions.[7]

More recently the international centres have devoted increasing amounts of their funds and manpower to outreach activities, defined by the Consultative Group on International Agricultural Research as 'those activities performed by centres away from their own research stations'. The original purpose of such activities was to test new materials for 'broad-based resistance to pests and disease, and to introduce these materials to farmers through national research, extension and production programmes'.

To what extent should the international centres become involved in national programmes of research, extension, and

[7] There are also many lands with ecosystems for which high-yielding varieties have not yet been discovered. A large proportion of these lands is in areas liable to uncontrolled flooding, typical of which are the heavily populated deltas of the Ganges, Bramaputra, Godavari, Cheo Praya, Irrawaddy, and Mekong rivers in Asia. Rice plants that are naturally adapted to flood conditions and which are at present grown in these areas include tall rices that stand out above the flood waters, certain rices that can survive for a few days under water, and the famous 'floating rice', which has the capacity to grow rapidly and keep abreast of the rising flood waters by being sustained on the surface by its buoyant leaves. Although there have been experiments to produce high-yielding rices of the submersible and also of the floating variety, no major success in these endeavours has yet been achieved.

Similarly, experiments continue in quest of high-yielding varieties for colder climates and, even more importantly, for unirrigated areas with irregular rainfall, where plants that thrive must be able to withstand prolonged dry spells. More than half the world's ricelands are of this 'upland' rainfed category.

An equally interesting and important set of problems for further research has to do with the modification of inhospitable soils and the identification or breeding of plant varieties that would thrive in them if they were suitably treated. Problems include acid sulphate soils (in perhaps 15 million hectares on which rice cannot be grown in Vietnam and Thailand); iron toxicity when certain soils are flooded (affecting 10–15 million hectares of rice soils), retarding growth and limiting yields; and zinc deficiency affecting 500,000 hectares in the Philippines and large areas of Pakistan and India. An area of 1,000,000 acres of such lands is reported to be situated only 10 miles outside Bombay. Should scientists discover packages making possible the use of these lands for rice growing, a great opportunity would be open to the government to settle landless and land-poor families as cultivators of low-cost medium or high-yielding rice or another suitable crop upon these soils. However, such a policy must be anticipated by the appropriation of such lands before the prospect of their reclamation initiates land speculation, leading once again to the adjournment *sine die* of the 'problem of hunger'.

production? May they not tend to weaken national efforts and even supplant national systems? The situation is complicated by a certain competition between international centres, and possibly between them and national governments, for scarce professional and technical manpower.

While there is little doubt about the usefulness of the international centres for the specialized research functions detailed above, as well as for providing certain research training facilities, it is very important that their policies should contribute everything possible to enable national research systems in agricultural science and technology to grow and prosper so that they acquire their own national capability to support, guide, and service national agricultural policy devoted to the solution of the problems we have been discussing.

International organizations should give careful consideration to the proposition that recommended technologies for a particular country should not be allowed to distance themselves dramatically from their existing technological capacity and infrastructural endowment. Moreover, reliance on external supplies and services should be supported by alternative self-provisioning and self-reliance arrangements, in case external sources of supply are cut off or become too onerous.

Nor is it necessarily the case that the research most essential for a country's development. is the most recent, the most sophisticated or the most difficult and expensive to carry out. There are many situations still in which the most significant results are to be obtained by the thorough application of well-known techniques and methods that can be easily learnt, and which may be much more consistently sustained than those that depend on constant reference back to sophisticated foreign research centres, and on the constant need for supplies from abroad.

It follows, therefore, that the most valuable assistance that can be provided by international organizations (in addition to long-term genetic-chemical research programmes) is that of helping countries to find and put into operation systems of research, extension, and production, using their own resources and characteristic approaches as far as possible, with reliance on foreign sources in exceptional circumstances. Meanwhile, long-term plans can be made for the improvement of research facilities and the training of scientific manpower. The most valuable technical assistance in research at any given moment is that which can make use of existing skills and is compatible with existing factor endowments and levels of infrastructure development.

However, for most countries, full research self-sufficiency may not be a goal, and for these countries, a further source of assistance has been experimented with and supported by FAO, namely the co-operation of networks of several research stations within a region comprising a number of countries. Working examples of such networks are to be found in the FAO Cereal Nursery programmes in the Near East and North Africa; the FAO Mediterranean and Near East Olive Improvement Project, the Inter-Asian and Andean Corn Programmes and the West African Rice Development Association. The examples given are all commodity-oriented but networks can also be factor-oriented (e.g. the FAO/FIAC fertilizer programme) or problem-oriented (FAO: Eighteenth Session, 1975).

V. NATIONAL RESEARCH SYSTEMS AND THEIR IMPORTANCE

The development of a scientific capability for agricultural research is a long-term task facing most of the newly independent nations and for this reason it should be given every possible support at the present time in order that the benefits may be enjoyed ten and twenty years ahead, when the problems faced by most of these countries are likely to be even more acute than they are today.

The importance of the international centres in the long-term development of genetic-chemical technology should not be allowed to relegate national research systems to second place, since it is not possible for a true national agricultural policy to be elaborated, maintained, and modified unless there are vigorous national centres, where contemporary scientific and technological knowledge can be brought to bear on the national problems with all their idiosyncracies and local specificities. The more obvious fields of competence that should at all costs be developed during the next decade include, in addition to the usual physical and biological sciences, farm management and agricultural economics, rural sociology, nutritional studies, survey and 'consultative research' skills, land and water utilization, soil conservation, farm systems with and without livestock, and practical engineering for the development of existing pumping, storage, transport, farm tools, and other middle-level technologies.

An attempt to plan an adequate national system of research was discussed by E. F. L. Abeyratne[8] in an article published in a Sri Lankan newspaper at the end of 1975. He recognizes not only the need for research centres dealing with the agronomic problems of the major climatic zones, but also the need for sub-centres in more limited areas:

Within each region there is still variability not only in physical conditions for agriculture but also in the socio-economic conditions of the farmers. The importance of providing research recommendations which can be adopted by farmers within the resources available to them cannot be overstated.

This requirement can be met by a further decentralization of the research programmes from the regional research station to the cultivators' fields. Such a programme of carrying out agronomic trials directly in cultivators' fields is the vital key to an effective system of agricultural research. It has a two-way relationship in that practical technical problems faced by cultivators are brought into focus on the one hand and the value of scientific research and new technology is brought home to the cultivator on the other. This system of cooperative experimentation between the research worker, the extension worker and the farmer is one of the most effective ways of popularizing science.

Dr Abeyratne's 'cooperative experimentation' is carried further and developed by Zandstra and colleagues working in the Caqueza Project in Colombia in the early 1970s in unpublished drafts and papers. The dialogue between cultivator and scientist is the central dynamic of what they call 'operational research', which, they say, should be an integral part of any programme of induced technological change and which they see as a process of trying out and jointly learning to apply new knowledge and materials.[9] Like Dr Abeyratne,

[8] Dr Abeyratne is Director of Research and Extension in the Department of Agriculture, Sri Lanka.

[9] A five-phase sequence of operations in the field is recommended. This begins with (1) the establishing of relations of confidence and mutual dependence with cultivators. The importance of this preliminary step still has to be stressed in view of the vanity of scientists of all kinds. Absence of relations of confidence has the

they insist that before recommendations arising from research about agronomic practices are propagated among cultivators, they must be adapted to the cultivator's access to resources and inputs in any given region.

The identification of the limitations and the formulation of prescriptions to overcome them (like experimentation with recommended practices) required the crystallizing of a 'research community' in which the cultivator participated with his local knowledge and practical experience. However, to give reality to this model of collaboration, the programme needed a capacity not only for experimentation but also for demonstration and popularization, so that new information about modification of structural limitation and research results could constantly be fed, in the rural vocabulary of the region, to the cultivators.

It is also to be assumed that elements of mobilization involved in the formation and activities of this research community will generate a certain amount of political weight, without which proposals for structural limitations are not likely to be implemented.

psychological effect of scrambling the information and undermining its credibility.

This is followed by (2) a careful study of the local production systems as they are in order to assess and compare them from the point of view of their contribution to the cultivator-family's livelihood. This operation can be simply performed on the basis of information on the cultivator's average net gain, the variability of his net income (to assess risk), the number of man-days used, and his total cash investment. Data can be examined for cultivators with high, middle and low levels of net gain.

With this information (3) the group proceeds to identify the existing potential for improving activities and methods, and these should then be presented to cultivators' families for checking and in order to broaden their understanding of their own system.

(4) Having understood the system of production in relation to the ecology and also to the socio-economic context, scientists and cultivators can propose experimental improvements.

The selection of subject matter for the experiments is done jointly and followed by a distribution of tasks to all those involved, with reference to the main research centre for supervision or technical assistance in carrying out the experiments. The experiments are conducted by cultivators on their own land, with control areas in the same farm cultivated according to local methods.

The final phase in the sequence of operational research, therefore, is (5) the identification of structural limitations on production systems and the separation of those that can be modified by shorter- or longer-term measures from those that cannot.

In outlining his ideas of a national system of agriculture, Dr Abeyratne proposes that the regional research centre within the country should form part of a larger but closely knit complex (which he later refers to as a 'regional agro-technical nerve-centre') containing also the extension service headquarters, as well as facilities for training staff and cultivators, and which can also serve as a base for experts in such subjects as water management, farm management and economics, crop production, horticulture, and so on.

The same realization of the need to bring scientific experiment and agricultural practice into fruitful interaction emerged in China, especially after the Cultural Revolution, as part of a policy of breaking down hierarchical distinctions between 'brain and brawn' and of avoiding the solidification of bureaucratic interests. This is how a scientist described the new situation to a recent visitor:

> Before the Cultural Revolution, we had twists and turns in our research methods. We trusted too much in specialists and technicians. We concentrated specialists in high-yield places such as Soochow and Chengtu and the Pearl River Delta—twelve places altogether. We attempted to have high-yield demonstrations and assumed that other places would learn. But we failed. We found that the specialists were not good at cultivating, not as good as peasants. (*Stavis, 1975*, p. 1971.)

Acceptance of the necessity of peasant participation has important implications for scientific research, which has hitherto been dominated by the systematic organization of experiments in various subject areas but focused upon a single commodity, such as rice, wheat or sorghum. However, this highly specialized research needs to be complemented by intensified research into place-specific farming systems. Such a change of direction is fully justified by the fact that most small cultivators are not monoculturalists but rely on a combination of lines of crop production, with possible supplementary activities in cattle-rearing, craftwork or fishing. Farm systems research is potentially a way of confronting the whole *problématique* of the small cultivator, which consists of organizing work and resources in order to obtain family livelihood by a feasible combination of self-provisioning and market operations.

The rejection by small cultivators of apparently productive

and profitable innovations may well be due to the difficulty of fitting them into the delicate balance of existing activities. When studied, farming systems frequently reveal an effective coming to terms with both natural environment and the socio-economic realities, and provide the optimum degree of security possible in the circumstances. However, the accumulated resources of experimentally gained knowledge, of modern methods of analysis, and of sophisticated apparatus available to contemporary science enable the scientist to contribute new and vital elements to these systems, provided their goals and constraints and the way they work in practice are fully understood.

Improvements in existing systems might be sought in several directions, but the determination of which direction is essentially a matter for the cultivator so long as the scientist is able to give him some idea of the range of possible improvements. Cultivators who rely on their own production for a large part of their family diet will welcome any contribution to their farming practices that makes possible the maintenance of an all-year-round food supply, whether it be by means of improved crop storage, off-season water supplies, or by the addition of a new line of production that is compatible with the existing system. Similarly, they may make an important breakthrough by adopting a change in timing and thus achieving a more regular all-year-round level of family labour use. Or perhaps the way forward may require the insertion of a cash-crop element into the system without undermining the self-provisioning element. One of the virtues of a varied combination of crops most appreciated by the small cultivator is its usefulness in risk-spreading.

Finally, it has many times been noted that tropical mixed economies are on occasions capable of achieving a very delicate balance in complementarities. There is no reason why the trial and error that produced these systems of ecological adaptation should not now be rendered more effective by low-cost or no-cost contributions from contemporary science.

XIII   PEASANT-BASED STRATEGIES

The possibilities for a government–peasant alliance are examined; and the key factor for the mobilization of working cultivators and the rural poor is identified as improved livelihood.

Common Features

Green Revolution strategy in most Asian countries has contributed to the increased economic polarization of rural populations and has not staunched the drift towards the marginalization and 'de-landing' of increasing numbers of working cultivators. However, in other Asian countries,[1] alternative development paths have been successful in promoting a steady advance in yields and cropping intensity while at the same time maintaining levels of rural employment and improving rural livelihoods.

China, Japan and Taiwan have all followed strategies that can be described as 'peasant-based' in the sense that they were able to carry the peasant majorities with them while

---

[1] At this stage, the discussion is narrowed down to the situation in the land-scarce tropical countries of Asia, where the new technology has the most immediate relevance. In Latin America, the historic bi-modal agrarian structure canalizes development capital into the large entrepreneurial farms, which continue to be attracted to highly profitable exports, while the peasant sectors, deprived of support for technological advances, tend towards relative decline, national food supplies become scarcer and more expensive food imports increase. The fate of various land reforms (Brazil, Chile, Colombia, Ecuador) and events in Cuba have shown that nothing short of a revolutionary change or a collapse of the power structure through military defeat could be expected to provide the necessary political basis for a peasant-based strategy.

The long-term prospects for most of the African countries are even more menacing for the great majority of cultivators, whose rights to cultivate land have hitherto been based on birthright through communal ascription. The great danger is that instead of finding ways of improving existing productive methods and structures, governments will permit the further alienation of communally held lands to urban-based and even external enterprises providing their own capital, and using more intensive or extensive mechanized cultivation methods. Although pressure on land is becoming acute only in a few areas, critical conditions could occur in a relatively short period if these methods are used for a rapid development either of export agriculture or of national food-producing agriculture.

definitely closing the door to a possible rival agricultural class constituted by a large commercial farm sector.

What were the economic factors that made possible the evolution of the mass of very poor peasant cultivators into efficient petty entrepreneurs and high-technology cultivators? And what historical circumstances brought into being the political will to follow a peasant-based strategy?

In the case of the earliest move into production involving bought inputs, that of Japan, transformation takes place against the background of a massive policy of 'Westernization'. As early as 1873, a Land Tax Revision Ordinance was introduced, imposing a tax of 30.5 per cent on net farm income to be paid in cash, and having the explicit purpose of 'shocking the peasants into commercial-mindedness'. Since cultivators were mainly tenants, this tax was additional to rents paid to the landlords. Gentler persuasion was also used and savings banks were established extensively. It is also reported that while these technological developments were taking place, the cultivators' cash position steadily improved with the rising price of the product in response to the pressure of demand as industrialization proceeded and as the towns grew.

Thus it is not surprising to learn that, by the 1930s, tenants and owner-cultivators in Japan were found to be able to save between $80 and $400 per year compared with a zero savings rate or even a minus one for small cultivators in most other Asian countries. India, for instance, had an annual rate of individual transferable cash savings of US$−1.3 in 1961/2 (*Ishikawa, 1967*).

The adoption of the new technology in Taiwan got under way between 1922 and 1939, after two decades of steady improvements in irrigation, infrastructure and husbandry. The substitution of chemical fertilizer for the various types of manure produced on the farm or in the village was a gradual one and distribution was managed by the state, which held a monopoly of chemical fertilizer, and operated through farmers' associations. Intensification of fertilizer application after the war was organized on the basis of barter, with different types of chemical fertilizers being repaid in paddy after the harvest, at ratios of equivalence established by the government. A good deal of discontent was expressed by the cultivators about these arrangements, and the government was considered to be overcharging.

The peasants also complained that they were obliged to accept fertilizer even when they did not wish to use it. Nevertheless, they received their fertilizer without having to borrow cash, and also without being obliged to repay loans by post-harvest sale when the price would normally be at its lowest.

China's adoption of high-fertilizer-response varieties is much more recent than Japan's and Taiwan's, getting under way during the 1960s. By 1972 it was calculated that China was using approximately one third the amount of fertilizer per hectare as was used in Japan. The Chinese policy was to concentrate the new technology in the areas designated 'of high and stable yields'; that is to say, areas in which land-and-water improvements had been completed and in which irrigation was mechanically controlled. Here the production and accounting unit is the work team, a farm consisting basically of the former land and the labour of some two dozen families. Work teams in areas of high and stable yields hold reserves enabling them to purchase for cash the chemical fertilizers they need. The work team is of sufficient magnitude to escape the economic fragility of the small cultivator with his individual plot, and in areas of high and stable yields is likely to be operating profitably. These collectives were in fact already 'progressive farmers', i.e. doing well enough to invest in and manage the new technology successfully.

The capacity of Taiwan, Japan, and China to maintain a peasant-based strategy and to modernize agriculture through their peasantries is accounted for by what happened to their political structures. In each of these three cases a significant rupture of the existing power structure occurred, displacing the rural landlord classes who had constituted village élites controlling local affairs.

The picture given by Dore (1959) of the relative strength of the poorer peasants (the majority of whom were tenants) in Japan during the 1920s is one of waxing strength accompanying technological development, union organization, and improving livelihood in the midst of growing conflict with the landlords. But this was followed by a deterioration in the 1930s as slump conditions due to the collapse of the silk market damaged agriculture and as an increasingly authoritarian government repressed the unions and their left-wing political supporters.

However, during the war itself, the peasants' vital strategic importance in feeding the nation as well as their importance as the main source of military manpower led the government to institute subsidies paid direct to the tenant to increase production, while there was a relative weakening of the landlord's position as *rentier* due to inflation.

Following the military defeat of Japan in World War II, land reform measures were introduced that effectively undermined the power and the economic position of the landlords, who had hitherto dominated village life. Tenants received security of tenure, controlled rents, technical and economic assistance, and facilities for instalment purchase of their land, which diminished rapidly in real value, so that most peasants were able to pay it off within a few years. The land reform measures were supported by large sectors of political opinion from liberal to communist. But the decisive political force ensuring that these radical measures were legislated and executed was that of the Supreme Commander of the Occupation, motivated by various reasons at that point in history to 'destroy economic bondage which had enslaved the Japanese farmer for centuries of feudal oppression', and contributed, it was said, to Japanese expansionism, by providing devoted soldiers and an economic need for foreign markets.

Thus, the Japanese peasantry came to occupy the central position in the agricultural sector by a sudden turn in the history of the country, which altered the power conjuncture.[2]

In the case of Taiwan, a similar exclusion of the landlords as a rival agricultural class took place, but under different circumstances. After Japan's defeat, Taiwan returned to Chinese sovereignty and was occupied and administered by

[2] Dore (1959) tells us that there were numerous cases of ex-feudal warriors (Samurai) who attempted to become entrepreneurial farmers, but that they met with no success. It is interesting to speculate on how the provisions of the land reform would have varied if they had been a growing class at the time of the Occupation.

a Kuomintang regime. In 1949, 1951, and 1953 a series of measures were enacted controlling rents payable by tenants, distributing state lands (formerly Japanese), and redistributing privately-held lands of more than three hectares. In this way a peasant-based structure was established that supported a high rate of growth in the national economy and itself rapidly increased productivity and diversified production. In the late sixties further extensive measures of consolidation of holdings, with the improvement of irrigation, drainage, and transport, were carried out by the state and charged to the individual landholders.

Was this the execution of a rationally conceived pro-peasant agricultural development policy? According to Apthorpe (1977), it cannot be so regarded: 'The leading objective of the Nationalist land reform at the end of the 1940s was to extinguish the threat posed to the immigrant regime by the emerging landlords. . . . The act was anti-landlord rather than pro-peasant.'

In both Japan and Taiwan, the social structure was shaken by external events in such a way that the full development of peasant agriculture became the most prudent policy for the government to follow. Given adequate technological and economic support, peasant agriculture offered stability and was more manageable than a class of large entrepreneurs with strong political connections.

In China, the sequence of policy stages leading to the present situation emerged from a more elaborated political philosophy, which prescribed transition from a peasant structure towards collective production units, and was politically feasible because the government that emerged was based on a politico-military alliance with the peasantry. This support that the communists were able to bring to government was based on years of leadership of the defence of peasant livelihood against both the foreign invader and the landlords, and the corrupt government that supported them. In the latter phases of the revolutionary process, the Party continued to generate the political support of peasant majorities by legitimizing the distribution of lands of the richer peasants and by making possible a steady improvement in the livelihood of the poorer peasant majorities.

It emerges from the three cases that, as a result of historic events larger than the agricultural sector and larger than the national framework of each case, the land-owning élites as rival contenders were excluded from direct competition in agricultural production. As a result of this exclusion, a peasant-based strategy has developed its own momentum in each case, even though the development paths and social systems are as distinct as are their historical antecedents.

The technological advance of productive forces has been generated in the agricultural sector by the transformation of peasant agriculture rather than by the imposition of entrepreneurial agriculture. Thus, the dynamic process of a fully capitalist development in the sector has either been eliminated or kept under control, and the marginalization and 'de-landing' of the peasant majorities, with all its social pathology, political instability, and individual trauma, has been avoided.

Can other countries of the Asian continent, or indeed of Africa and Latin America, promote an adequate technological development by adopting peasant-based strategies? The determination of the development policies of each country is in principle its own affair, and depends on a unique set of changing conditions. However, in order to facilitate the struggle for a development path that is not disastrously discriminatory against the poor majorities, some further elements are combined with the common features of the three countries examined in order to present a normative picture of a 'peasant-based strategy'.

Two essential features for the operation of a peasant-based strategy are (i) access by agricultural families to land on the basis of relative equality assured by property ceilings or by a rational division of agricultural lands among work teams; (ii) facilitation of the growth of the productive potential of farms on the basis of shared capitalization rather than by monopolistic cumulative appropriation, which we have called the talents effect.

In China, saving and investment may take place at the level of the family backyard economy, of the work team itself, of the brigade, the commune or the county, according to the conveniences of scale of the particular item of investment. In

Japan and Taiwan, farmers' associations and co-operatives become repositories of the savings of the individual farm and invest in technical services, processing plant, and social facilities. In this way, trends towards polarization are slowed down.

Actual control of the talents effect depends on further administrative action by the state. In Taiwan particularly, state services to peasant producers, coupled with heavy taxation, have played an important role. In China, special measures to limit inequalities between communes in richer and poorer areas have been taken. In Japan, there seem to be persistent pressures towards uneven accumulation that are not in fact controlled. However, by the 1970s the questions affecting the rural producer have already been absorbed into the much larger *problématique* of any modern industrial society moving towards automation. In other countries (e.g. India and Malaysia) the magnitude and orientation of government investment in the agricultural sector shows signs of a partial commitment to shared capitalization, yet the very existence of profitable opportunities for investment in larger-scale commercial or entrepreneurial agriculture by individuals and corporations with access to capital continually weakens and subverts programmes in which investment is directed at the common interests of the working cultivator, through mechanisms that have already been illustrated in earlier chapters.

The fact is that, in a market society, unless there is persistent political pressure for the continuation of the strategy, it can be reversed quite quickly, and the whole impetus of growth may pass to an entrepreneurial sector, articulating itself rapidly out of the larger peasant-producers and into the service entrepreneurs, ex-landlords, and investors from outside the sector.

The most spectacular case of this kind was that of Mexico, described earlier. Tunisia followed the same path. Certainly the principle of shared capitalization in agriculture requires a political will for its consistent application, which goes sharply against the competitive principle of the free market ideology; and it cannot be based on land tenure alone, as the *ejido* experiment in Mexico has shown.

The evidence also suggests that without the occurrence of

some sort of *bouleversement*—one that results in a government committed to excluding social formations other than the peasantry from any leading role in direct agricultural production—a successful peasant-based strategy is not feasible. But in fact the likelihood of convulsive change and rupture for most of the poor countries is great, in any case.

Elements of instability include international warfare; the further disorganization of world trade by the breakdown of the present international financial arrangements; internal strife involving ethnic rivalries and state power; and the threat to national food systems based on self-provisioning constituted by the ever-growing demands of the rich world on the produce of the land of the poor countries—and a resultant outbidding of the poor for basic food supplies.

The kinds of crises envisaged are such that the moments when policy options are open become more frequent. In these circumstances, the United Nations may have a specially valuable role to play in making available to nations and governments technical services and information about alternative strategies from other countries' experiences more appropriate than the models proffered by the great powers, which the governments of poor countries are under pressure to adopt.

A government that adopts a peasant-based strategy is likely to look to the peasantry for a large part of its political support and must be able to mobilize that support both in achieving power and in maintaining it. One of the first principles of this alliance is an agricultural development policy that not only fulfils certain national requirements but also ensures the palpable improvement of the livelihood of the working cultivator, so that livelihood becomes the prime measure of success of the policy. In the first instance, this must mean generalized food security and an end to malnutrition and the danger of famine.

In other words, the keynote for mobilization of the working cultivators and rural poor becomes 'improved livelihood', with increased production aimed at as both cause and consequence, *but not as primary aim*. And the success of rural and agricultural development programmes then becomes manifest in such indicators as children's age-to-weight ratio

and in higher levels of participation in development programmes.

Given large-scale relative equity, shared capitalization, and
the exclusion of entrepreneurial agriculture, the continuing
politics of the peasant-based strategy can be most aptly seen
as a dialectical interaction between state power and peasantry
—a mobilizing action on the part of the state concerned with
national development goals, the agricultural surplus, and its
own political support, and a variable participatory action by
the peasantry.

Great importance must be given to peasant movements
since those who set out to put into practice a peasant-based
strategy must look to such movements for their support in
obtaining and retaining power, along with the support of
other social forces, and for both purposes are likely to develop
programmes of peasant mobilization by which effective use
can be made of political forces that have manifested themselves hitherto only on a local scale.

Serious conflicts and problems requiring careful analysis
are likely to arise in the process of mobilizing peasantries.
Many of the movements of rural poor that achieve a certain
scale of political operations owe an important element of
their force and solidarity to common ethnic, linguistic, and
religious identity different from that of the power-holders of
the national state.

A particularly sensitive aspect of the mobilization of
peasant movements is their leadership. As well as representing certain common community interests, such leadership
is frequently exercised above all in respect of minority
(e.g. ethnic) grievances, although the leaders themselves are
likely to belong to that class of the peasantry that is interested
more in cheap labour and individual capital accumulation
than in shared capitalization. The leadership of peasant movements, for example, is frequently that of the aspiring entrepreneurial farmer and service entrepreneur. For this reason
the concept of 'working cultivator' is a key one for the
analysis of peasant mobilization and of the constitution of
existing movements.

Following the adoption of a true peasant-based strategy,
relations between peasantry and the state become those of an

asymmetrical partnership. The progress of peasant agriculture is essential to the state and the state is not likely to admit other important contenders to the play. But the whole system of facilities mounted by the state to improve technology, facilitate marketing, and provide inputs acquires the characteristics of monopoly, and the administration of such services leads to the creation of a powerful rural bureaucracy with its hands on the flow of wealth, goods, and influence between working cultivator and the larger society from which agricultural supplies and consumers come.

Whether the peasant-based strategy operates through individual family farms or through co-operative or collective farms, new types of exploitation and abuse may well emerge, either through excessive demands by the state made on the peasant producer through price policy, grain deliveries, or taxation, or by the greed of the new rural officials.

Consequently, two good reasons are offered for giving great importance to the participatory organization of peasantries, or more precisely, working cultivators and labourers. Future conjunctures in which a peasant-based strategy would become official policy are unlikely to occur without the activities of organizations and participatory movements becoming a recognized political force in the polity, whether attached to supporters of the government or exercising power in opposition as a following of political personages or as an illegal trade union or insurrectionary movement. Many such movements have drawn peasants into the national liberation struggles against colonial powers or ethnically oppressive governments. They can produce solidarities transcending purely local action by peasants against particular abuses.

The character of participatory movements after the institution of a peasant-based development strategy is equally important and equally difficult. It is necessary to defend the strategy itself from internal abuse, from exaggerated state extortion for investment in other sectors and from the emergence of bi-modality, namely, agrarian structures in which agricultural production units (farms) divide into two clearly distinguished strata in respect of the magnitude of the farms as economic undertakings, the lines of cropping, their market orientation, and the socio-political situation of the cultivators who control them.

Whatever its use, the abstract elaboration of a desirable peasant-based strategy can soon reach the limits of plausibility. From this point on, further work has to be conducted in the context of real situations where such strategies might feasibly have a future.

But a question still remains as to the usefulness of research. We have stressed the low level of manoeuvrability and freedom of choice enjoyed in policy decision-making by governments. What impact can research findings be expected to have on social and economic measures affecting the lives of people? The answer is that research output can have an important influence if it is available in the right form, at the right time, and in the right place.

The right place is any significant 'policy arena', or point of encounter where policies are fought over and adopted. If policies are considered to be 'any sustained course of action adopted and pursued by a group for its advantage' then it is both legitimate and desirable to look for policy arenas far beyond and below the high councils of government. The party caucus, the junior officers' club, the transnational boardroom, the peasant union executive, the diplomatic cocktail party, the group of student activists, and many semi-institutionalized cliques constitute such arenas. Clearly, the results of field research must be widely disseminated in different modes of readability for them to penetrate these arenas.

Unfortunately, the net effect of much research is to throw an enigmatic veil over the true character of social and economic problems by hiding them in academic virtuosity and obscurantism. Social research must play a more positive role by revealing social realities in the light of universally humane values. Without departing from just observations, it can serve in the mobilization of political will and political power for the solution of the problems so revealed.

PUBLICATIONS AND DOCUMENTS CITED IN THE TEXT

AMERASINGHE, N., *Social and Economic Implications of New Varieties of Rice: Case Study of Minipe Colony*, a Global-2 Report, in *Hameed, 1977* (see below).

APPU, P. S. 'The Bamboo Tubewell', *Economic and Political Weekly* (Vol. IX, no. 26, 29.6.1974).

APTHORPE, RAYMOND, *The Burden of Land, An Asian Model Land Reform Re-analysed*, Institute of Social Studies 25th Anniversary Conference, The Hague, December 1977.

BANDYOPADHYAY, SURAJ, 'Impressions of What is Changing Under the Impact of the Community Development Blocks Along the Two Sides of the River Mayurakeshi' (mimeo), Indian Statistical Institute, Calcutta, 1972.

BAPNA, S. L. 'Social and Economic Implications of the Green Revolution: A Case Study of the Kota District', a Global-2 Report (mimeo), Agro-Economic Research Centre, Vallabh Vidyanagar, India, 1973.

BARKER, RANDOLPH, 'The Evolutionary Nature of the New Rice Technology', paper presented to Conference of Japanese Economic Research Centre, Tokyo, IRRI, Los Baños, Philippines, 1971.

—— 'Labor Absorption in Philippine Agriculture' (mimeo), IRRI, Los Baños, Philippines, October 1972.

——, DOZINA, G. Jr., and FU-SHAN, LIU, *The Changing Pattern of Rice Production in Japan, Nueva Ecija, 1965 to 1970*, IRRI, Los Baños, Philippines, 1971.

——, MEYERS, W. H., and CORDOVA, V., *The Impact of Devaluation on Fertilizer Use and Profitability in Philippine Rice Production*, IRRI, Los Baños, Philippines, April 1974, cited in *Palmer, 1975* (see below).

BARRACLOUGH, SOLON L., 'Rural Development Strategy and Agrarian Reform', paper presented at Latin American Seminar on Agrarian Reform and Colonization, Chiclayo, Peru, 1971.

BARTH, G. F., 'Economic Spheres in Darfour', in Firth, Raymond (ed.), *Themes in Economic Anthropology*, Tavistock Publications, London, 1970, cited in *Feldman and Lawrence, 1975* (see below).

BARTSCH, WILLIAM H., 'Employment Effects of Alternative Technologies and Techniques in Asian Crop Production', a Global-2 Report (mimeo), UNRISD, Geneva, 1972.

BELL, C. L. G. and S. D. PRASAD, 'The Condition of Share-croppers in Purnea District', (mimeo) 1972.

BERNAL, E. A., 'The Role of Landlords in Philippine Agricultural Development', M.A. thesis, University of the Philippines, 1971, cited in *Castillo, 1975* (see below).

BETEILLE, ANDRÉ, 'Caste, Class and Power: Changing Patterns of Stratification in a Tanjore Village' (mimeo), Berkeley, 1965.

BHALLA, G. S. *et al.*, *Changing Structure of Agriculture in Haryana (A Study of the Impact of the Green Revolution)*, Punjab University, Chandigarh, 1972.

BHATI, U. N., *Some Social and Economic Aspects of the Introduction of New Varieties of Paddy in Malaysia: A Village Case Study*, UNRISD Report No. 76.8, Geneva, 1976.

BURKI, SHAHID JAVED, *A Study of Chinese Communes*, Cambridge, Mass., 1969.

CARR, MARILYN, 'Some Social and Economic Aspects of Tractorization in Ceylon' (mimeo), January 1975.

CARRUTHERS, I. D. and CLARK, COLIN, *The Economics of Irrigation*, Liverpool University Press, 1980.

CASTILLO, GELIA T., *All in a Grain of Rice*, Southeast Asian Regional Center for Graduate Study and Research in Agriculture, Laguna, Philippines, 1975.

CENTRO DE INVESTIGACIONES AGRARIAS (CIDA), *Estructura Agraria y Desarrollo Agricola en México*, 3 vols., Mexico City, 1970.

CHAMBERS, ROBERT, Paper presented at a seminar on problems of rice-growing areas of Tamil Nadu and Sri Lanka, *Centre of South Asian Studies*, 1974.

CHAUDHARI, HAIDER ALI, and ABDUR RASHID, 'Changes in Rice Farming in Gujranwala, West Pakistan' (mimeo), West Pakistan University, Lyallpur, 1972.

— and QAMAR MOHY-UD-DIN, 'Changes in Rice Farming in Gujranwala, Pakistan', paper presented at IRRI Conference, Los Baños, Philippines, 1973.

CHINNAPPA, B. NAJAMMA, 'Rice Cultivation in Tamil Nadu— Strategy, Achievements and Constraints' (mimeo), University of Madras, Madras, 1972.

CHOWDHURI, B. K., and G. OJHA, 'A Study of High Yielding Varieties Programme in the District of Saran, Bihar, with Reference to Hybrid Maize (Kharif) 1968–69' (mimeo), Agro-Economic Research Centre, Santiniketan, India, 1969.

COLLIER, SOENTORO, and G. WIRADI, 'Rice Harvesting and Selling Changes in Central Java which Have Serious Social Implications', *Agro-Economic Survey*, Research Notes No. 14, 1973.

DALRYMPLE, DANA, *Development and Spread of High Yielding Varieties of Wheat and Rice in the Less Developed Nations*, U.S. Department of Agriculture, Washington DC, 1974 (I).

— *The Green Revolution, Past and Prospects*, USAID, Washington DC, 1974 (II).

DANDA, AJIT K., and DIPALI G. DANDA, *Development and Change in Basudha*, National Institute of Community Development, Hyderabad, 1971.

DASGUPTA, BIPLAB, *Agrarian Change and the New Technology in India*, UNRISD Report No. 77.3, Geneva, 1977.

DESAI, D. K., 'Intensive Agricultural District Programme', *Economic and Political Weekly*, Review of Agriculture, Bombay, June 1969.

DESAI, GUNVANT M., *Increasing Fertilizer Use in Indian Agriculture*, Cornell University, Ithaca, 1969.

DORE, R., *Land Reform in Japan*, London, 1959.

DUEWELL, JOHN, and OSMAN BIN MOHAMED NOOR, 'Socio-economic Survey of Tenancy Patterns in Trengganu Paddy Production' (manuscript), March 1971.

DUMONT, RENÉ, *Notes sur les implications socialies de la 'Révolution Verte' en quelques pays d'Afrique*, UNRISD Report No. 71.5, Geneva, 1971.

EASTER, WILLIAM K., 'Neglected Opportunities in Irrigation', *Economic and Political Weekly*, Vol. IX, No. 14, Bombay, 6 April 1974.

ECKSTEIN, SALOMON, *El ejido colectivo en México*, Fondo de Cultura Económica, Mexico, 1966.

FAO, *Provisional Indicative World Plan for Agricultural Development: Summary and Main Conclusions*, Rome, 1970.

— *Monthly Bulletin of Agricultural Economics and Statistics*: Vol. 22, No. 11, November 1973. Vol. 23, No. 10/11, November 1974. Vol. 24, No. 2, February 1975.

— *Production Yearbook 1971*, Rome, 1971 (I).

— *Introduction and Effects of High-Yielding Varieties of Rice in the Philippines*, Rome, 1971 (II).

— *Production Yearbook 1972*, Rome, 1973 (I).

— *Production Yearbook 1973*, Rome, 1974.

— 'Report of the Special Committee on Agrarian Reform' (mimeo), Rome, November 1971 (III).

— *The State of Food and Agriculture 1972*, Rome, 1972.

— *The State of Food and Agriculture 1973*, Rome, 1973 (II).

FELDMAN, DAVID, and PETER LAWRENCE, 'Africa Report', a Global-2 Report (mimeo), UNRISD, Geneva, 1975.

FRANKE, RICHARD W., 'The Green Revolution in a Javanese Village', Ph.D. thesis (mimeo), Harvard University, 1972.

FRANKEL, FRANCINE, excerpt from 'Agricultural Modernization and Social Change', in review article by P. C. Joshi in *Seminar*, May 1970.

GEZA, SAM, *The Socio-economic Implications of the Large-Scale Introduction and Adoption of High-Yielding Maize Varieties in Zambia: A Case Study of the Kwenje and Pemba Areas*, a Global-2 Report, Lusaka, 1975.

GÓMEZ, MAURO, 'Plan Puebla: Avances en Seis Años de Operación' (mimeo), paper presented at International Seminar on Rural Regional Development Projects, Bogotá, September 1972.

GRIFFIN, KEITH, *The Green Revolution: An Economic Analysis*, UNRISD Report No. 72.6, Geneva, 1972.

HALE, SYLVIA, *Barriers to Free Choice in Development*, VIII World Congress of Sociology, Toronto, 1974.

—— 'Barriers to Free Choice in Development: A Case Study of Four Uttar Pradesh Villages', *International Review of Community Development* (Rome), No. 33–4, 1975.

HAMEED, N. D. ABDUL, 'Tenancy and Resource Use in Peasant Agriculture—A Case Study' (mimeo), Institute of Development Studies, University of Sussex, 1971.

—— *Rice Revolution in Sri Lanka*, UNRISD Report No. 76.7, Geneva, 1977.

HART, GILLIAN, 'Patterns of Household Labour Allocation in a Javanese Village' (mimeo), Department of Agricultural Economics, Cornell University, February 1977.

HAURI, IRÈNE, *Le Projet céréalier en Tunisie: Études aux niveaux national et local*, UNRISD Report No. 74.4, Geneva 1974.

HAVENS, EUGENE, and WILLIAM FLINN, 'Green Revolution Technology: Structural Aspects of its Adoption and Consequences', a Global-2 Report (manuscript), UNRISD, Geneva, 1975.

HEWITT DE ALCANTARA, CYNTHIA, *Modernizing Mexican Agriculture: Socio-economic Implications of Technological Change 1940–1970*, UNRISD Report No. 76.5, Geneva, 1976.

HOPPER, D., 'Investment in Agriculture: the Essentials for Pay-off', *Strategy for the Conquest of Hunger*, Rockefeller Foundation, n.d.

HUNTER, GUY, *The Administration of Agricultural Development—Lessons from India*, Oxford University Press, London, 1970.

—— and ANTHONY F. BOTTRALL, *Serving the Small Farmer: Policy Choices in Indian Agriculture*, Croom Helm/ODI, London, 1974.

ISHIKAWA, SHIGERU, *Economic Development in Asian Perspective*, Tokyo, 1967.

—— *Agricultural Development Strategies in Asia: Case Studies of the Philippines and Thailand*, Asian Development Bank, 1970.

—— 'A Note on the Choice of Technology in China', *Journal of Development Studies*, vol. 9, October 1972, pp. 161–86.

JACOBY, ERICH H., 'Extension-credit—Input Delivery', a Global-2 Report (manuscript), UNRISD, Geneva, 1973.

—— 'The "Green Revolution" in China', a Global-2 Report (mimeo), UNRISD, Geneva, 1974.

JIMENEZ, LEOBARDO, 'Strategies for Increasing Agricultural Production on Small Holdings: the Puebla Project', 'Conference of Food Research Institute, Stanford University, 13–16 December 1971.

JEQUIER, Article in *OECD Observer* (No. 75, May/June 1975).

JOSHI, P. C., 'Agrarian Social Structure and Social Change' (mimeo), Institute of Economic Growth, Delhi, 1971.

KAHLON, A. S., A. C. SHARMA, and P. C. DEB, 'Organizational Implications of Rapid Growth and Commercialization of Punjab Agriculture'. (In *Hunter and Bottrall, 1974*—see above.)

—— and GURBACHAN SINGH, *Social and Economic Implications of Large-Scale Introduction of New Varieties of Rice in the Punjab with Special Reference to the Gurdaspur District*, a Global-2 Report, Punjab Agricultural University, Ludhiana, 1973 (I).

KAHLON, A. S. and GURBACHAN SINGH, *Social and Economic Implications of Large-Scale Introduction of New Varieties of Wheat in the Punjab with Special Reference to the Ferozepur District*, a Global-2 Report, Punjab Agricultural University, Ludhiana, 1973 (II).

KERKVLIET, B., *Political Change in the Philippines*, Honolulu, 1974.

KHUSRO, A. M., *Economics of Land Reform and Farm Size in India*, Macmillan Co., Delhi, 1973.

KRISHNA, DAYA, *The New Agricultural Strategy*, Rajkamal Press, New Delhi, 1971.

LAXMINARAYAN, H., *Social and Economic Implications of Large-Scale Introduction of New Varieties of Wheat in Haryana: Part I. General; Part II, Statistical Statements*, a Global-2 Report, Agricultural Economics Research Centre, Delhi, 1973.

MAIR, LUCY, *Studies in Applied Anthropology*, London, 1957. Cited in *Feldman and Lawrence, 1975* (see above).

MANGAHAS, MAHAR, AIDA E. RECTO, and VERNON W. RUTTAN, *Production and Market Relationship for Rice and Corn in the Philippines*, IRRI, Los Baños, 1970.

— and AIDA LIBRERO, 'High-Yielding Varieties of Rice in the Philippines, a Perspective', a Global-2 Report (manuscript), UNRISD, Geneva, 1974.

MEARS, LEON A., and MELIZA H. AGABIN, 'Finance and Credit Associated with Rice Marketing in the Philippines' (mimeo), University of the Philippines, 1971.

— and TERESA L. ANDEN, *Rice Marketing Costs and Margins*, Institute of Economic Development and Research, University of the Philippines, 1972.

MEILLASSOUX, C., *Anthropologie économique des Gouro de la Côte d'Ivoire*, École Pratique des Hautes Études, Paris, 1970.

MELLOR, JOHN W., and T. V. MOORTI, 'The Dilemma of State Tube-Wells', *Economic and Political Weekly*, Vol. VI, No. 13, Bombay, March 1971.

MENAMKAT, A., 'Development Problems and the Role of Credit Co-operatives in Indian Agriculture', *Ph. D.* dissertation, University of Fribourg, n.d.

MENCHER, JOAN P., 'Change Agents and Villagers: Their Relationships and Role of Class Values', *Economic and Political Weekly*, Vol. V, Nos. 29–31, Bombay, July 1970.

— 'Conflicts and Contradictions in the Green Revolution: the Case of Tamil Nadu', *Economic and Political Weekly*, Annual Number, Bombay, February 1974.

NARAIN, D., and P. C. JOSHI, 'Magnitude of Agricultural Tenancy', *Economic and Political Weekly*, Review of Agriculture, Bombay, September 1969.

NERFIN, M., *Entretiens avec Ahmed Ben Salah sur la dynamique socialiste dans la Tunisie des années 1960*, François Maspero, Paris, 1974.

PALMER, INGRID, *Science and Agricultural Production*, UNRISD Report No. 72.8, Geneva, 1972.

— *The New Rice in the Philippines*, UNRISD Report No. 75.2, Geneva, 1975.

— *The New Rice in Asia: Conclusions from Four Country Studies*, UNRISD Report No. 76.6, Geneva, 1976.

— *The New Rice in Indonesia*, UNRISD Report No. 77.1, Geneva, 1977.

PARE, LUISA, 'Two Villages in the Puebla Plan: Santa Isabel Tepetzala and San Andrés Hueyacatitla', a Global-2 Report (manuscript), UNRISD, Geneva, 1972.

PARTHASARATHY, G., 'Changes in Rice Farming and their Economic and Social Impact: Case Study of a Delta Village, Andhra Pradesh, India' (mimeo), Waltair, Andhra Pradesh University, 1973.

PEARSE, ANDREW, 'Agrarian Change Trends in Latin America', *Latin American Research Review*, Vol. I, No. III, 1966.

— *The Latin American Peasant*, Frank Cass, London, 1975.

PIMENTEL, DAVID *et al.*, 'Food Production and the Energy Crisis', *Science*, Vol. 182, No. 4111, Washington DC, 2 November 1973.

POLEMAN, THOMAS T., and DONALD F. FREEBAIRN (eds.), *Food, Population and Employment: the Impact of the Green Revolution*, Praeger, New York, 1973.

PREBBLE, JOHN, *The Highland Clearances*, Secker and Warburg, London, 1963.

RAJ, K. N., 'Mechanization of Agriculture in India and Sri Lanka', *International Labour Review*, Vol. 106, No. 4, ILO, Geneva, October 1972.

RAY, S. K., 'Imbalances, Instability and Government Operations in Foodgrains', *Economic and Political Weekly*, Bombay, September 1970.

RIVERA, SALOMON, 'Las Implicaciones sociales y económicas debidas a la introducción de semillas mejoradas de alto rendimiento: a Pradera', a Global-2 Report (manuscript), Bogota, 1972.

— 'The New Agricultural Technology in the Cauca Valley, Colombia', a Global-2 Report (manuscript), UNRISD, Geneva, 1972 (Résumé by Antonio Barreto).

RYAN, JAMES, RUPERT SHELDRAKE, and SATYA PAL YADAR, 'Human Nutritional Needs and Crop Breeding Objectives in the Semi-Arid Tropics' (mimeo), 1974.

SAJOGYO, 'Modernization without Development in Rural Java' (mimeo), FAO, Rome, 1973.

SCHLUTER, MICHAEL, *Differential Rates of Adoption of the New Seed Varieties in India: the Problem of the Small Farm*, Cornell University, Ithaca, 1971.

SCHOFIELD, S., 'Village Nutrition' (manuscript), 1974.

SCHULTZ, THEODORE, *Transforming Traditional Agriculture*, New York, 1968.

SELVADURAI, J., *Social and Economic Implications of New Varieties of Rice in Sri Lanka: a Case Study of Palannoruwa Village*, a Global-2 Report. In *Hameed, 1977* (see above).

SELVANAYAGAM, S., *Social and Economic Implications of New Varieties of Rice in Sri Lanka: a Case Study of Palamunai Village*, a Global-2 Report. In *Hameed, 1977* (see above).

SINGH, ROSHAN, 'Social and Economic Implications of Large-Scale Introduction of New Varieties of Wheat in Muzaffarnagar', a Global-2 Report (manuscript), Agra, India, 1973.

STAVIS, BENEDICT, *Making Green Revolution: the Politics of Agricultural Development in China*, Cornell Rural Development Committee, Ithaca, 1975.

STEINHART, J. S., and C. E., 'Energy Fuel in the U.S. Food System', *Science*, Vol. 184, No. 4134, Washington DC, 19 April 1974.

TAKAHASHI, AKIRA, 'Peasantization of Kasama Tenants: Socio-Economic Changes in a Central Luzon Village', paper presented at IRRI Seminar, 1971.

TAUSSIG, MICHAEL, 'Peasants and the Expansion of Sugar Production in Cauca Valley' (manuscript), 1971.

THOMAS, JOHN W., 'The Choice of Technology in Developing Countries: the Case of Irrigation Tubewells in East Pakistan', *LTC Newsletter*, University of Wisconsin, July–September 1973.

UNITED NATIONS RESEARCH INSTITUTE FOR SOCIAL DEVELOPMENT (UNRISD), *The Social and Economic Implications of Large-Scale Introduction of New Varieties of Foodgrain: Summary of Conclusions of a Global Research Project*, UNRISD Report No. 74.1, Geneva, 1975.

— *Strategy and Programme Proposals*, UNRISD (1977) W.P.2/Rev.1.

UNITED STATES DEPARTMENT OF AGRICULTURE, *Changes in Agriculture in 26 Developing Nations*, Washington DC, 1965.

— *Agricultural Development and Expansion in the Nile Basin*, Washington DC, 1968.

— *The Agricultural Situation in the Far East and Oceania*, Washington DC, 1972.

VAN DER KLOET, HENDRIK, *Inégalités dans les milieux ruraux: Possibilités et problèmes de la modernisation agricole au Maroc*, UNRISD Report No. 75.1, Geneva, 1975.

VOHRA, B. B., *A Charter for the Land*, Ministry of Agriculture, New Delhi, 1972.

VYAS, V. S., 'Tenancy in a Dynamic Setting', *Economic and Political Weekly*, Review of Agriculture, Bombay, June 1970.

—, 'Rural Works in Indian Development', *Development Digest*, Vol. XI, No. 4, Washington DC, October 1973.

WANG, SUNG-HSING, and RAYMOND APTHORPE, *Rice Farming*

in Taiwan: Three Village Studies, Academica Sinica, Taipei, 1974.

WEINTRAUB, LEON, 'Introducing Agricultural Change: the Inland Valley Swamp-Rice Scheme in Sierra Leone', Ph.D. thesis (mimeo), University of Wisconsin, 1973.

WERTHEIM, W. F. and L. CH. SCHENK-SANDBERGEN, *Polarity and Equality in China*, Voorpublikatie 7 (1973), Universiteit van Amsterdam, Reprint, 1976.

WHITCOMBE, ELIZABETH, 'Notes on Indian Irrigation Policy', contributed by the author, 1974.

WINKELMANN, DON, 'Plan Puebla after Six Years', paper presented to Ford Foundation Seminar, Mexico City, 6–10 November 1972.

— 'Factors Inhibiting Farmer Participation in Plan Puebla', *LTC Newsletter*, No. 39, University of Wisconsin, Land Tenure Center, January–March 1973.

ZANDSTRA, HUBERT G., *Metodología para la comparación de estratégias en la recommendación de fósforo*, Bogota, April 1973.

INDEX